Hope and his Edinburgh contemporaries – after John Kay

1. The Rev Hugh Blair **2.** Benjamin Bell **3.** Alexander Monro, *secundus* **4.** James Burnett, Lord Monboddo **5.** James Hutton **6.** Joseph Black **7.** William Cullen **8.** The Rev John Walker
9. Francis Home **10.** Andrew Duncan **11.** John Hope **12.** Malcolm McCoig **13.** Alexander ('Sandy') Wood **14.** James Gregory **15.** James Bruce of Kinnaird **16.** Sir William Forbes
17. The Rev William Robertson **18.** Andrew Bell **19.** William Smellie **20.** Henry Hume, Lord Kames **21.** Hugo Arnot **22.** Sir Archibald Hope **23.** James Hope, 3rd Earl of Hopetoun

Hopea odorata

John Hope

(1725–1786)

Alan G. Morton's
Memoir of a Scottish Botanist

A new and revised edition

Hugh Noltie

H.J. NOLTIE

Royal Botanic Garden Edinburgh

MMXI

For Paul Nesbitt
in recollection of our first Hope exhibition of 1986
and in celebration of 25 years of exhibitions in Inverleith House

Set in Baskerville, in commemoration of Hope's meeting
with John Baskerville in Birmingham, 1766

First edition by A.G. Morton published by the Edinburgh Botanic Garden (Sibbald) Trust, 1986
This edition published by the Royal Botanic Garden Edinburgh, 2011
20A Inverleith Row, Edinburgh EH3 5LR
www.rbge.org.uk

Royal
Botanic Garden
Edinburgh

ISBN 978-1-906129-71-2

Designed by Caroline Muir, RBGE
Printed by Scotprint, Haddington, Scotland

Frontispiece. *Hopea odorata*. Hand coloured
engraving by Daniel Mackenzie, after a
drawing by an anonymous Indian artist,
published in William Roxburgh's *Plants of
the Coast of Coromandel* (1811).

Contents

Foreword

This new and greatly enlarged edition of A.G. Morton's *John Hope 1725–1786: Memoir of a Scottish Botanist* is to be welcomed for several reasons. First, and most importantly, since its first publication in 1986 much new information has become available concerning John Hope, an important figure in the botanical history of Scotland and, indeed, the world. As Henry Noltie explains, much of this new material has come to light as a result of the Botanic Cottage project, which has systematically dismantled the gardener's house that stood on the former site of the Royal Botanic Garden at Leith Walk with the hope of resurrecting it in the Inverleith garden. Another source of new insights has been a close inspection of the lecture notes taken by some of Hope's students, now preserved in the Garden's archives. Secondly, in naming the RBGE's new western entrance on Arboretum Place the John Hope Gateway, my distinguished eighteenth-century predecessor has inevitably been the focus of fresh interest and attention.

Whilst Professor Morton's book included a selection of Hope's correspondence (omitted in the present edition), Henry Noltie provides a much fuller account of Hope and his connections within Edinburgh society at the time of the Scottish Enlightenment. His colleagues and acquaintances are illustrated as they were portrayed at the time by John Kay, the Scottish caricaturist and engraver. Hope himself might have been a relatively low-key figure, leading 'a simple and unostentatious' life, but his influence as a teacher of botany was profound. A parallel can be drawn with his slightly earlier contemporary, the great Swedish botanist Carl Linnaeus, who was greatly admired by Hope, and who famously sent his botanical disciples far and wide around the world in search of new discoveries. The catalogue of acclaimed botanists taught by Hope is remarkable, and includes such names as John Lindsay, Archibald Menzies, William Roxburgh, Sir James Edward Smith and William Wright.

As with all of Henry Noltie's books, this one is rich in fascinating insights. From 'the notation by two different students of Hope's pronunciation of the word "grasses" as "cresses"', he deduces that Hope appears to have 'had the refined Edinburgh diction later associated with the suburb of Morningside'. I know few other historical writers who would conjure the character of their subject's voice from archival materials dating back well before the earliest sound recordings. Other personality traits are interpreted from Hope's reaction to his head gardener's, John Williamson's, death at the hands of armed smugglers, and the monument he erected to his memory. The book provides both a fuller understanding of John Hope, a significant Scottish scientist, and a wider perspective on a key period in Scottish history. I am sure that you will enjoy reading it as much as I have.

Stephen Blackmore

Professor Stephen Blackmore CBE, FRSE
Regius Keeper and Queen's Botanist

Reviser's preface

In 2009 a major new building was opened at the Royal Botanic Garden Edinburgh (RBGE) – a physical and intellectual 'Gateway', combining orientation, educational, retail, and catering functions. A public competition to name the building resulted in its denomination as 'The John Hope Gateway', in commemoration of the distinguished Enlightenment-period Regius Keeper of the Garden. This, together with an exhibition at Inverleith House in 2011 of some of the remarkable large-scale drawings made to illustrate Hope's botanical lectures, seems an appropriate time to publish a new, more amply illustrated, edition of † Professor Alan Morton's biographical memoir published on the occasion of the bicentenary of Hope's death in 1986.

Since then much new material has become available, some the result of exciting studies based around the archaeological survey, controlled demolition (with the intention to rebuild within the present garden at Inverleith) of Botanic Cottage, the only relic of Hope's great Leith Walk garden remaining *in situ* until 2008. One of the most surprising results has been the discovery that Botanic Cottage was far more than the head gardener's home and an entrance to the garden from Leith Walk – such had always been known – but it emerges that the upper room was where Hope delivered his lectures, and that, despite the humble appearance of its latter days, it was almost certainly designed by no less an architect than John Adam. So our new Gateway by a leading contemporary architectural practice is far closer to Hope's, in spirit and function, than was realised or intended, and history has unwittingly repeated itself.

Given the outstanding richness of the documentation for the Leith Walk garden – in terms of its building history and maintenance, as much as of the plants grown – it seems incomprehensible that no study has ever been made of this unique research and teaching establishment. It is similarly extraordinary, given a plethora of studies of the Scottish Enlightenment (including Edinburgh's renowned medical faculty and of his colleague the Rev John Walker), that the fate of Hope's contribution to that intellectual world, and wider botanical history, has continued to be that of passing mentions and relegation to footnotes. Research into both of these topics – his garden and his lectures – is currently in progress, and the present reviser of Morton's text is working on an edition of the botany lectures, so that Hope's thought, research and teaching can be seen in their rich and wide-ranging detail for the first time. The original edition of Morton's book will remain in print, as it includes items omitted from the present one (notably translations of Hope's letters to Linnaeus, and notes on taxonomic and nomenclatural matters written by the late B.L. Burtt).

† Alan Gilbert Morton (1910–2003) was educated at the Universities of Liverpool and Cambridge, obtaining a PhD at the latter on carbohydrate metabolism in ivy leaves in 1937. His career was spent largely in applied botanical and mycological research, including appointments with Lever Brothers and ICI, before he became Professor of Botany at Chelsea College, University of London, in 1966. He retired in 1973 and moved to Edinburgh where he edited the *Transactions of the British Mycological Society*, and in 1981 published his valuable *History of Botanical Science*. A gifted linguist, in both ancient and modern languages, his major non-botanical interests were in poetry and philosophy.

The life and work of John Hope

In the botanical world of the second half of the eighteenth century, from Virginia to Calcutta, by way of Madrid, Paris, Aleppo and St Petersburg, Professor John Hope of Edinburgh was well known and respected for the 'more modern and scientifick appearance' he imparted to the study and teaching of botany, and for the skill and energy with which, in little more than twenty years, he built up the Royal Botanic Garden Edinburgh to be one of the leading institutions of its kind in Europe. The geographical spread of his renown came in a large measure from the widely various origins of the medical students, more than 1700 of whom attended his lectures over a period of a quarter of a century, and their subsequent equally wide dispersal, which was not restricted to the then expanding British colonies. For these reasons it is worth providing an account of the labours and achievements of this neglected botanist of the Scottish Enlightenment, whose characteristic concerns of 'improvement, sociability, humanity, toleration, and intellectual cultivation' he shared and exemplified.

Hope was born in Edinburgh on 10 May 1725, the son of Robert Hope, a surgeon, and his wife Marion Glass of an ancient family of Sauchie, Stirlingshire. The family to which John Hope belonged was descended from Henry Hope, an Edinburgh merchant of the sixteenth century, through his elder son Thomas (1573–1646), who became a successful advocate and distinguished legal expert (from James, a younger son of the merchant, derived the Amsterdam banking branch of the family, with whom Hope retained contact). Like many lawyers Thomas Hope made large quantities of money from his profession, purchased an estate in Fife, and ended up Sir Thomas Hope of Craighall. Appointed King's Advocate by Charles I, Sir Thomas nevertheless stoutly defended both Presbyterians and the Covenanters against the English king and court. Besides five daughters, Sir Thomas's wife bore him nine sons, giving rise to a not inconsiderable clan. Many members of the family prospered, especially in the legal profession or in public office, prudently investing savings in the purchase of landed property or in proto-industrial enterprises such as lead- and coal-mining. They were solid citizens who mostly supported the

Fig. 1. Dr John Hope (left) talks with his gardener (probably Malcolm McCoig). Etching by John Kay, 1786.

'Glorious Revolution' that brought William & Mary to the throne in 1689, and later accepted the Hanoverian succession.

John Hope was the great-great-grandson of Sir Thomas Hope through his eldest son another John (1605–1695, 2nd Baronet of Craighall). The earls of Hopetoun were descended from Sir Thomas's seventh son James and Dr John Hope acted as physician to the second earl and his lady, an appointment doubtless due to cousinly relationship at four removes. Hope's grandfather, Archibald Hope of Rankeillor (1639–1706), was a younger son of Sir John Hope of Craighall, who became an eminent advocate and made himself unpopular with James VII (II of England) but under William & Mary was knighted and appointed a Judge of Session with the title Lord Rankeillor. Robert Hope, John Hope's father, was the youngest of Lord Rankeillor's five sons. It is unfortunate that our knowledge of Robert Hope is so limited: he is recorded as entering the Incorporation of Chirurgeons in 1716 and as its president in 1720–1. In his 1789 obituary of John Hope, Andrew Duncan recorded that Hope's father was a surgeon of great ability and extensive practice. Robert Hope was clearly also a man of generosity and public spirit, for in 1729, a year after the opening of the Royal Infirmary, he was, with Alexander Monro *primus*, one of six 'Chirurgeon-Apothecaries' who offered 'to attend the Infirmary in

their turns without any reward or salary, to dispense the medicines prescribed by the Physicians faithfully from their shops ... and generally perform the duties of chirurgeons and apothecaries to the Infirmary' (Peel-Ritchie 1899).

The young John Hope was sent not to Edinburgh High School (as would have been usual for the son of an Edinburgh professional), but to the grammar school of Dalkeith, where one of his fellow pupils was John, elder son of the architect William Adam. At Dalkeith Hope acquired a good grounding in the learned languages, and in later life was to correspond in tolerable Latin with Linnaeus and other foreign botanists, and he was even consulted on the identity of Greek plants by Lord Monboddo (Knight 1900). From school Hope chose to follow his father's example and seek a career in medicine. Records of his entry into Edinburgh University and attendance at classes have been lost, but he would have begun his medical studies at about the age of fifteen, probably in 1740 or 1741. Possibly he attended lectures by the great mathematician Colin Maclaurin as part of his premedical course, for Maclaurin also taught the elements of Newtonian natural philosophy (physics). The name of John Hope (presumably our one) appears in a list of those attending the medical class of Alexander Monro *primus* in the years 1743 and 1746. Whether Hope's studies were interrupted during the brief occupation of

Edinburgh by the Jacobite army in the autumn of 1745 is not known. Hope was elected a member of the Medical Society of Edinburgh in April 1745 and in December 1748 received honorary membership of the Society 'by succession' (i.e., automatically by seniority), which exempted him from regular attendance at meetings. About this time he followed in the footsteps of many newly qualified medical men, and went abroad to gain additional experience and instruction in medical schools and hospitals in Europe. John Hope chose Paris rather than Leyden, the traditional resort of Edinburgh graduates, because, interested in botany from an early age, he wished to study under Bernard de Jussieu, Demonstrator in the Jardin du Roi, whom, according to Duncan, and presumably based on personal information, Hope 'had already long admired'.

John Hope's early attachment to botany may owe something to his father, who as a surgeon-apothecary would have had a professional interest in plants and herbs, including those of the Scottish countryside, and who must surely have introduced his son to two of the main physic gardens then existing in the city – the Royal Abbey Garden at Holyroodhouse, and the Town Garden at Trinity Hospital. Thomas Hope, Robert's elder brother, followed the family tradition of the law; he inherited (indirectly) the Craighall baronetcy, but was notable as a keen agricultural improver. He was responsible for draining the eastern end of the Meadows, still known in consequence as Hope Park. Sir Thomas lived to a great age and perhaps also helped to form the botanical bent of his youthful nephew, especially its applied aspects. Sir Archibald Hope (grandson and heir of Sir Thomas as 9th Baronet of Craighall and Pinkie. EP22) sponsored a student to attend the 1775 botanical lectures of his cousin-once-removed, and was probably also partly responsible for John Hope's interest in racing (the only pastime of which we have incidental record) – for this 'Knight of the Turf' was one of the main supporters of the Leith Races.

Hope's enthusiasm for the science of botany was certainly fired by the teaching of Charles Alston (1683–1760), King's Botanist and Professor of Botany and Materia Medica in the medical faculty of Edinburgh University, whose classes Hope necessarily attended. Alston had trained in medicine at the University of Leyden under the illustrious Hermann Boerhaave, an enthusiastic and distinguished botanist. When Hope was his student, Alston as King's Botanist had already been keeper of the Royal Physic Garden at Holyroodhouse since 1716, but had been appointed University Professor (and in charge of the Town Garden at Trinity Hospital) only in 1738. His published writings prove Alston to have been a capable botanist with a thorough knowledge of the literature of the subject, but his memory has suffered because of his opposition to the

Linnaean reform of plant nomenclature and his rejection of the sexual function of pollen in seed formation. John Hope had no doubt of the quality of his teacher or of his own indebtedness to him – as shown by the fact that in 1770, ten years after Alston's death, he undertook the arduous labour of editing for publication his teacher's lectures on materia medica, in two hefty volumes.

It seems impossible to recover, as one would so much like to do, details of John Hope's visit to Bernard de Jussieu in Paris, a visit that had great influence on his botanical thinking and activity for the rest of his life. Only three documents survive as evidence of the Hope-Jussieu connection. The first is a small, undated scrap bearing a note (in Latin) of Antoine de Jussieu's opinion of J.P. de Tournefort's method of classification. Since Antoine (Bernard's elder brother) died in 1758 and left few published remains, this appears to be a contemporary memorandum by Hope of a conversation that must have taken place at the Jardin du Roi in 1748/9. The other two documents testify to the continuation of Hope's links with Paris. In 1765 Bernard de Jussieu sent Hope his compliments via 'Mr Burnet', almost certainly the advocate James Burnett, later Lord Monboddo, who is known to have been in Paris in 1765 in connection with the notorious Douglas case. Hope responded on 24 September 1765 with a letter to 'my esteemed teacher and benefactor'. This letter is preserved in the Jussieu

autograph collection in the Muséum National d'Histoire Naturelle in Paris, and in it Hope asked Jussieu to send him an account of 'the arrangement of the Plants which you have made at Trianon', which probably accounts for the third relevant document, among Hope's own papers, titled 'Methodus Plantarum horti regii Paris', with a ground plan of 'A Class in Jussieu's Ecole botanique'.

The length of Hope's French sojourn, and any other places visited on the only foreign journey he ever undertook, is unknown. If he was present at the 1748 Medical Society meeting in Edinburgh, then his time in Paris must have lasted less than a year, for there is no doubt that he returned to Scotland at some point in 1749. According to Andrew Duncan, Hope's return from Paris was precipitated by his father's death but this cannot be correct, since the death of Robert Hope, surgeon in Edinburgh and surgeon to the garrison in the Castle, is recorded in the *Scots magazine* of January 1742, when Hope was at an early stage in his medical studies. In January 1750 Hope obtained a doctorate of medicine at the University of Glasgow. Having already qualified by examination in the prescribed medical courses, to gain an MD degree Hope would only have had to present and defend a thesis on a selected topic, though the subject of his thesis does not appear in the Glasgow records. As with many other aspects of Hope's life prior to 1761, the reason for his choosing to graduate from Glasgow is unknown. Was it merely a

Fig. 2. Edinburgh Royal Infirmary, designed by William Adam in 1738, demolished 1884. Engraving after a drawing by the Hon. John Elphinstone, published in Arnot's *History of Edinburgh* (1779).

formality, as implied by Morton, or did he in fact spend some time at Glasgow University? Though often forgotten, it had a physic garden at this time and Robert Hamilton held the joint Regius Chair of Botany and Anatomy. Little is known of Hamilton's botanical teaching, if any, but from 1747 to 1750 William Cullen gave botanical lectures associated with this physic garden, and for the first time in Scotland taught Linnaean classification. It is an intriguing possibility that Hope might have attended these. The reason for taking an MD at this juncture was doubtless to better his career prospects, but Hope could just as easily have taken his degree at Edinburgh, suggesting that there was a particular reason for taking

it at Glasgow. Might this also have been connected in some way with the Stevenson family into which he later married? In any case Hope returned to Edinburgh where, in November 1750, he became a member of the Royal College of Physicians and thereby licensed to practise medicine.

For the next ten years Dr Hope was a busy and successful medical practitioner in Edinburgh, noted for his care of and humanity towards his patients. He continued to treat private patients all his life, which provided a part of his income that he greatly valued; but in 1768 he was appointed Physician to the Royal Infirmary [fig. 2], where, up to his death, he gave equally

humane and conscientious treatment to the poor and needy. In 1762 he was elected Fellow of the Royal College of Physicians of Edinburgh and in 1784 became its president [fig. 3]. He played an active part in moves by the College to induce the Town Council to take measures to improve the health and cleanliness of the city (Craig 1976).

In what ways the young Dr Hope kept his botanical interest green in these early years, when fees from attending patients were his only income, is uncertain: he must certainly have kept abreast with botanical literature and perhaps found time to collect plant specimens for his herbarium while making visits outside the city. He was also drawn into those circles of intense intellectual activity that became the Scottish Enlightenment and made Edinburgh a centre of European culture. A characteristic feature of this movement was its exchange of ideas through clubs and debating societies, so it is not surprising that Hope

belonged to several of these in addition to the Medical Society. One of the most influential of these debating clubs was the Select Society, started by the painter Allan Ramsay in 1754, with a fluctuating life over a ten-year period. According to Emerson (2004) the initial membership consisted of 15 lawyers, clerics and academics, among whom were the philosopher David Hume, Adam Smith (Professor of Ethics at Glasgow), James Burnett (advocate, later Lord Monboddo. EP4), Francis Home (physician, pioneering agricultural chemist and later Hope's successor as Professor of Materia Medica. EP9) and the

physician Alexander Stevenson, shortly to become Hope's brother-in-law. In May 1755 the initial group was augmented to 30, and among the new members were John Hope, Hugh Blair (minister of Lady Yester's Church, later Professor of Rhetoric and Belles-lettres. EP1) and William Robertson (minister of Gladsmuir, later Principal of Edinburgh University. EP17). Several men who were of importance in Hope's life soon joined the growing Society: the medics William Cullen (EP7), Alexander Monro *primus* and Sir Alexander Dick, the lawyer Henry Home, Lord Kames (EP20), the minister the Rev Dr John Walker (later Professor of Natural History. EP8) and the banker Adam Fairholm. The Select Society met weekly to debate any subject under the sun 'except such as regard Revealed Religion or which may give occasion to vent any Principles of Jacobitism', but specialised in 'socio-economic subjects' (Emerson 2004). It also spawned two splinter groups, to both of which Hope belonged – the Edinburgh Society for the Encouragement of Arts, Science, Manufactures and Commerce, and the Select Society for Promoting the Reading and Speaking of the English Language in Scotland. The former awarded numerous financial prizes ('premiums') on improvement subjects, from clover seed to

Fig. 3. Physicians' Hall, George Street, designed by James Craig, built 1775, demolished 1844. Wooden model. The building in which Hope presided between 1784 and his death. Royal College of Physicians of Edinburgh.

book binding, and Hope must surely have taken an interest in the offer of a premium for a discourse on the principles of vegetation, a proposal mentioned by David Hume in a letter to Allan Ramsay in 1755. On matters of concern to the second of these societies it is of interest to note that in 1782 Hope borrowed Diderot's *Œuvres de théâtre* from the University Library. And from the notation by two different students of Hope's pronunciation of the word 'grasses' as 'cresses', it would appear that he had the refined Edinburgh diction later associated with the suburb of Morningside. Of bodies that catered for the more 'scientific' interests of Hope and many of his fellow members of the Select Society, he also belonged to the Philosophical Society (Emerson 1981, 1985). Membership of this society included scientists and men of influence from all over Scotland, united in a desire to apply science to the 'improvement' of the nation; David Hume (with Alexander Monro *secundus*. EP3) was its joint secretary from 1751 to 1763.

Hope was thus known to the leading figures in the public and intellectual life of Edinburgh. From his student days his somewhat reserved but generous disposition, and the principled motives that manifestly guided his activity, enabled him to gain and preserve the lifelong regard of many friends. One of these from the days of the Select Society was the eccentric Monboddo. For many years Hope was a regular guest at Monboddo's classical suppers, held each week while the Court of Session was sitting, and at which Joseph Black (Professor of Chemistry at Edinburgh. EP6) and James Hutton (founder of scientific geology. EP5) were frequently present, as was the learned printer William Smellie (EP18), who will be discussed later.

After becoming established in practice, on Sunday 24 February 1760, Dr Hope married Juliana Stevenson, daughter of the late Dr John Stevenson who had been an eminent Edinburgh physician. As bride and groom were both residents of New North Kirk parish the wedding must have taken place in the nave of St Giles, then walled off from the rest of the mediaeval church. The marriage contract reveals that while Hope brought £300 to the union, he married an heiress – Juliana brought a dowry of £1700 sterling. She also apparently brought some of her father's botanical books, as Hope's copies of the works of Robert Morison bear John Stevenson's bookplate. At this time (certainly in 1765) Hope appears to have been living in Craig's Close, on the north side of the High Street immediately to the east of John Adam's Royal Exchange (now the City Chambers). In the year of Hope's marriage the venerable Charles Alston died, an event that brought a turning point in Hope's career.

Involving two posts (and two gardens), the Crown, the Town Council and the College of Surgeons, the politics of the appointment of Alston's successor were likely to be complex, though the benefits of vesting the responsibilities in one man were by this time clear. In fact the process went smoothly and there seems to have been no rival candidate. Emerson (1988) has explained the background – involving the patronage of Archibald Campbell, Duke of Argyll, at the end of his life and a career that involved, among much else, the overseeing of Crown appointments in Scotland. Emerson cites the appointment of Hope in April 1761 as King's Botanist for Scotland and Superintendent of the Royal Garden, and by the Town Council as joint Professor of Botany and Materia Medica in the Faculty of Medicine in the University, as the first such appointment attributable to Argyll's nephew John Stuart, 3rd Earl of Bute [fig. 4]. Hope himself considered that 'it was to his Lordship that I owed my office' and was forever grateful to Bute. The new professor was well qualified in both his subjects by the thorough, if in some respects old-fashioned, training under Alston, whilst in botany he was conversant with the foremost developments and ideas as a result of his experience at the Jardin du Roi and from the close study of recent botanical literature revealed in his lectures. Botany was from this time the main business of Hope's life, although he never relinquished his medical career, private and public, and seventeen years later could state that his practice was still increasing. Hope's role in the establishment of the garden at Leith Walk, and his botanical teaching, will be discussed in separate chapters.

For the first seven years of his joint appointments, Hope lectured on Materia Medica in the University during the winter, and on Botany at the Botanic Garden from May to July. However, his health began to suffer under the strain of running two university courses, a botanic garden, and a medical practice. He was by now known as a leading botanist – both nationally and internationally. He had published on *Rheum* in the *Philosophical transactions of the Royal Society* and had been elected a Fellow of the Royal Society in 1767. He was sufficiently influential with the Edinburgh Town Council, patrons of the University, to persuade them to separate the teaching of Botany from that of Materia Medica. Two new Regius Chairs were created, one for Materia Medica, which Hope first offered to David Skene who declined, so it went instead to Dr Francis Home. The Chair for Medicine and Botany Hope himself retained. At the same time the office of King's Botanist and Superintendent of the Royal Garden, which Hope already held 'at the King's pleasure', was conferred on him for life, and an annual salary of £50 attached to the office. So John Hope became in 1768 the first Regius Professor of Botany in the University of Edinburgh, and henceforward his academic responsibilities were limited to teaching and research in what he called his 'beloved Science'.

Fig. 4. John Stuart, 3ʳᵈ Earl of Bute (1713–1792). By Allan Ramsay, 1758.
Private collection at Mount Stuart.

At the same time his financial rewards were materially improved. The annual salary of a Regius Professor was £77 sterling, to which was added £50 as King's Botanist. Far more substantial, however, were the course fees he received from students. Each student paid two guineas for the Materia Medica course; the fee was the same for Botany until 1771 when it was raised to three guineas. Hope was particular about payment, with the benefit to historians that he made lists of every student who attended his lectures from 1761 to 1786, and whether or not they had paid. He did on occasion reduce the fee for the many who took his course on more than one occasion, but his parsimony in such matters was noted by at least one student. Against this, he did allow a number of free places each year to fellow professors and their families, influential Edinburgh citizens, divinity students, surgeons' apprentices, and protégés of his aristocratic friends (though doubtless he received payment at least in kind for some of these). It was not all take, however, and each year he entertained to supper (again, meticulously listed), presumably at his own house, all the gentlemen taking the Botany course: in 1780 there were 59, invited weekly in groups of six while the course was in progress. Nonetheless, Hope made a handsome income from his academic posts: for example, for the year 1762 he had 55 Materia Medica and 58 Botany students, so received £235 in fees, giving, with his royal and academic salaries, a total annual income of £362, which excludes the unknown and doubtless substantial income from his private medical practice. Using conversion rates based on the retail price index, this is about £50,000 in today's terms; by comparison, the annual salary of a contemporary schoolmaster defended in a legal case by James Boswell was £20, whilst the annual salary of a judge of the Court of Session was then £700.

Although he had chosen to free himself from the burden of lecturing on Materia Medica, Hope could not, of course, neglect medicinal ('officinal') plants in the teaching of Botany to medical students, as already seen in his editing of Alston's lectures. The resulting care he took of officinals in the Leith Walk garden is shown in the detailed instructions on their arrangement and spacing written out for his gardeners, and the list of them he published – anonymously in two, undated, editions (Appendix 1). As a member of the Royal College of Physicians Hope was also involved in preparing the editions of the *Pharmacopoeia* published periodically by that body. Two editions were published during Hope's period – a sixth in 1774 and a seventh in 1783. His involvement in the latter is certain, as he consulted both Sir Joseph Banks and Linnaeus's son over it.

In his later years Hope was an active and prominent figure in civic, academic and medical affairs, as President of the Royal College of Physicians (December 1784 until his death), Principal Physician to the Royal Infirmary, a governor of the Orphan Hospital, and a member of the University Senate.

In 1773 he was a founder member of the Aesculapian Club, inaugurated by Andrew Duncan, which met on Fridays in various taverns under the patronage of Apollo, Bacchus and Venus 'to promote goodwill and fellowship between physicians and surgeons' (Chalmers 2010). In 1778 Hope was a founder member of the Newtonian Club, an offshoot of the Philosophical Society, and in 1783 he took part with William Cullen (since 1773 Professor of the Practice of Physic) and William Robertson (by now Principal of the University) in drawing up a memorial to the Crown, which led to the founding of the Royal Society of Edinburgh,

of which Hope was thus a foundation Fellow. By the mid 1770s Hope and his family were living at an address variously given as 'Head of High School Wynd' and 'High School Yards', probably at the junction of the two, opposite Lady Yester's church, and conveniently close both to the High School (which his sons attended) and to the Royal Infirmary [fig. 5].

Fig. 5. High School Wynd looking north from the gates of the old Royal High School. Hand coloured lithograph by James Drummond, published in his *Old Edinburgh* 1879, based on a drawing made by Drummond, c. 1848.

The house in the foreground, at the junction of High School Yards, is probably where Hope lived (the building with the hexagonal corner turret is Cardinal Beaton's house on the Cowgate).

Fig. 6. The west wall of Greyfriars kirkyard, etching by Daniel Lizars, c. 1790. Hope is buried in front of the blank section of wall below the dormer window in the low building far right. The triangular-pedimented wall monument to the left of this is that of his great-great-grandfather, Sir Thomas Hope of Craighall.

the element strontium, leading to a move back to Edinburgh to assist Joseph Black, to whose chair in Chemistry he succeeded in 1799. Thomas Charles Hope was to hold this chair for 44 years, and was noted for his spectacular, and spectacularly well attended, lectures in what is now the Talbot Rice Gallery of the University.

It was in the full vigour of his many activities, three months after finishing his annual Botany course, that John Hope died on the evening of Friday 10 November 1786, of a sudden illness that baffled the medical knowledge of the time. He was buried close to the west wall of Greyfriars kirkyard [fig. 6], in the family burial ground of his cousin Sir Archibald (Anderson 1931). Surprisingly no contemporary stone was raised in this enclosure, its extant inscriptions being the self-aggrandising Victorian-period ones on unweatherable Rubislaw granite of Hope's lawyer son James Hope WS and his immediate family, with no tribute to his distinguished antecedents, and only a lateral mention of his distinguished academic brother Thomas Charles. James's son, the philanthropist John Hope, did, in 1866, make amends in Old College with a pair

John and Juliana had five children. The eldest, Robert (born 1761) must have died relatively young (though not before fathering a daughter), as it was Robert's younger brother John (1765–1840), a Major in the Royal Irish Rifles, who inherited much of their father's property. The other children were a daughter Anne Marion (1763–1837), who married James Walker of Dalry, and two further sons: James (1769–1842) who continued the family's legal tradition, and Thomas Charles (1766–1844). Of these Thomas was the most interesting: at Edinburgh University he was something

of the perpetual student, attending his father's lectures on no fewer than five occasions. His MD thesis was on the motion and life of plants, and he aspired to succeed his father. Probably for largely political reasons, as much as to his youth, this did not happen, but through the influence of his uncle Alexander Stevenson, Professor of the Practice of Medicine at Glasgow, he was in 1787 appointed to lecture in Chemistry and Materia Medica in the western university, and succeeded to the medical chair on the death of his uncle in 1791. He held this position for four years, during which time he discovered of drinking fountains at the mid-points of the long sides of the quadrangle. These once functional memorials (understated to the point of near invisibility in the magnificence of their architectural context) are dedicated to his botanist grandfather and chemist uncle. Recently, thanks to the endeavours of Muriel Ann Hope (who married a descendent of Sir James Hope of Hopetoun and took a great interest both in family history and the Leith Walk garden project), the deficiency at Greyfriars was remedied and a plaque to Hope dedicated in 2004. When vandals broke this, her sons replaced it in 2008 with a stone more firmly rooted in the earth, which, it is to be trusted, will mark Hope's final resting place for posterity.

Hope left his family comfortably off: from the 'testament dative and inventory' of the goods at his death we know of 'Twelve thousand pounds Scots [= £1000 sterling] of Capital stock in the hands of the Governors and Company of the Bank of Scotland'; but there must have been much more than this, as this does not include moveable property (such as his valuable library), or any real estate (such as the land of, and surrounding, the Leith Walk garden).

Those who knew Hope all testified to his humane and kindly disposition, to his unselfish and unremitting devotion to public duty and the promotion of knowledge, and to his scrupulous integrity.

Apparently Hope confided to his intimate friends that his temper was naturally irascible and could only be controlled by constant effort of the will (perhaps the source of the headaches and stomach pains mentioned in an astonishingly frank memoir on his own health written in 1777 – literally down to pains in his gonads and sweaty feet at times of anxiety). No doubt his sound good sense did not allow him to suffer fools gladly; but if he did sometimes lose his temper this never seems to have disturbed the cooperative relations he maintained throughout his life with men in every rank of society. According to Andrew Duncan, his style of living was simple and unostentatious, which enabled him to help promising young men struggling against poverty: one recorded instance is his generosity (jointly with James Robertson, Professor of Oriental Languages) in providing William Smellie with £70 to set up in the printing business. As will be seen later it also allowed him to launch the medico-botanical careers of Archibald Menzies and John Lindsay, and the anatomical-artistic one of Andrew Fyfe.

A few snatched contemporary glimpses may be added to fill in the above bare outlines of Dr Hope's active and fruitful, but largely uneventful, life. A young American student of medicine, Benjamin Rush, a protégé of Benjamin Franklin, confided to his diary in 1766 that 'Dr Hope, Professor of Botany, is a man of little genius but great application, his virtues as a Gentleman have procured him the love of every body'. James Boswell, as a busy advocate at the Scottish Bar, knew 'the worthy and ingenious' Dr Hope well and in 1784, concerned about Dr Johnson's failing health, he wrote to Hope and two other eminent Edinburgh physicians, Cullen and Monro, for advice. All three sent courteous replies, and Hope responded by writing to his old friend Dr Brocklesby in London, then a near neighbour and intimate visitor of Dr Johnson, whose physician he was. In replying to Boswell, Hope wrote: 'few people have better claim on men than your friend as hardly a day passes that I do not ask his opinion [via his *Dictionary*] about this or that word'. Whether Hope made Dr Johnson's acquaintance during one of his visits to London is not known, though it is a possibility in view of their common friendship with Sir John Pringle. One meeting between Hope and Johnson was, however, recorded by Boswell. When Johnson was returning to London on 22 November 1773, after his famous tour of the Highlands with Boswell, he boarded the coach at Blackshields, about twelve miles south of Edinburgh. Hope happened to be on the coach, travelling as far as Newcastle, and he and Dr Johnson were travelling companions as far as that city. According to Boswell they each in later years used to recall 'their good fortune in this accidental meeting, for they had much instructive conversation'.

Hope and the Leith Walk garden

One of John Hope's greatest and most lasting contributions to the advancement of botany was his transformation of the Edinburgh Botanic Garden as a permanent teaching and research institute of a new type, inspired by the ideas of Joseph Pitton de Tournefort, and the brothers Antoine and Bernard de Jussieu, as enshrined in the Jardin du Roi in Paris, and which in Hope's time came to be known as the 'Royal Botanic Garden'.

On his appointment in 1761 Hope set about a project for creating a new botanic garden in Edinburgh to replace the Royal Abbey Garden and the Town Garden at Trinity Hospital, which were too small for development and badly affected by city smoke. The Town Garden on marshy ground (now occupied by Waverley Station) was also threatened by the planned building of North Bridge. Hope proposed to unite the two gardens on a single larger site outside the city, and chose a plot to the north of Leith Walk that would be free of pollution, but within relatively easy reach of the University; its soil turned out to be rather poor – of clays, sands and gravel – but a marsh at its centre was ideal for turning into a pond. Alston had tried several times without success to get government funding for a proper botanic garden.

It has already been seen that Hope owed his position to Lord Bute [fig. 4], whom he referred to in a letter to Linnaeus as a 'Maecenas' (a patron of learned men and letters). Bute's patronage also took a more personal form, as seen in five valuable books he gave to Hope now in the RBGE library. Bute's scientific and botanical interests had been fostered by his uncle Argyll, and flourished through his association with Frederick, Prince of Wales, and his wife Princess Augusta, at Kew. Bute's botanical and taxonomic interests (and his support of John Hill, an author much quoted by Hope) require more attention than they have yet received (along lines started by Miller 1988). These interests were manifest in Bute's collection of plants at his Robert Adam mansion, Luton Hoo in Bedfordshire, and more especially in his published *Botanical tables*, an empirical and strictly pragmatic attempt to group plants together by similarity in a belief that a truly natural system was an unattainable goal, but in groups that were more robust than those of Linnaeus. After Frederick's death in 1751 Bute's advisory role on developing the gardens at Kew continued until the death of the Dowager Princess Augusta in 1772 (when he was replaced by Banks), as did his mentoring role over the young

Prince George who became king on the death of his grandfather George II in 1760. Bute's political career was a blighted one that has only recently been reassessed (Schweizer 1988), but Hope shrewdly took advantage of his brief tenure as Prime Minister (May 1762 to April 1763) to gain further patronage over funding for his new garden. Hope's astute methods and timing had their reward and he received a grant from the Lords of the Treasury in 1763 so that the new botanic garden could be set up.

Though not initially intended, and due to the collapse of the finances of the Fairholm family to whom the land belonged, it appears that the property side of the deal turned into something of a 'private finance initiative', as Hope himself purchased a 999-year lease on a 13-acre plot of land on which the 5-acre botanic garden was to be laid out, and which he had John Leslie survey in January 1763. The details of the lease are unclear and were the subject of legal disputes in the nineteenth century, but Hope (and later, rightly or wrongly, his heirs) was paid an annual ground rent of £25 for the land occupied by the garden, and doubtless the intention was to build on the surrounding land, which eventually started in the 1820s.

Hope's was an ambitious project and in 1763 he asked his old school friend, the architect John Adam [fig. 7], to draw up an estimate for 'making a botanic garden 4 Scots acres or 5 English'. Adam's finances were tied up with the Fairholm banking enterprise, suggesting that Hope's commission involved a mixture of motives – aesthetic, practical and financial, and perhaps partly explaining Hope's purchase of the site (Adam Fairholm jumped from a boat while fleeing to Europe from his creditors, so Hope probably got a bargain). John Adam's estimate was for an enclosing six-foot-high drystone wall, a gardener's house, a greenhouse and two hothouses, and a hotbed, which, with levelling and trenching of the site, came to £1502; with the annual upkeep reckoned at £112. Presumably this is the sum Hope applied for: the Treasury actually gave him £1330 1s 2½d for making the garden, and £69 annually for its upkeep; the Town Council agreed to pay the annual ground rent. So in December 1763 Hope could write informing the College of Surgeons of his success, having taken care to ensure their support in the previous year when his application to the Crown was pending. The funds granted enabled Hope to establish the new garden on Leith Walk, and John Adam (or at least his office) is almost certainly responsible for the design of the gardener's house and the impressive 140-foot-long suite of conservatories with its central 70-foot greenhouse, linked to a 26-foot hothouse at both eastern and western ends. For the plans of these buildings Adam was paid £15 on 29 March 1765. When Hope needed to extend the East Stove in 1779, and for improvements to the garden's frontage, he went to another eminent architect, James Craig, designer of Edinburgh's New Town and of

Fig. 7. John Adam (1721–1792). By Francis Cotes. Mr & Mrs Keith Adam.

the hall of the College of Physicians on George Street (now occupied by 'The Dome' bar and restaurant) [fig. 3].

The rarer plants from the two former sites were transferred (those from the Town Garden being purchased from Alston's widow Bethia for £21 sterling in May 1765) and efforts made to acquire new plants both from within Scotland and from Hope's network of foreign

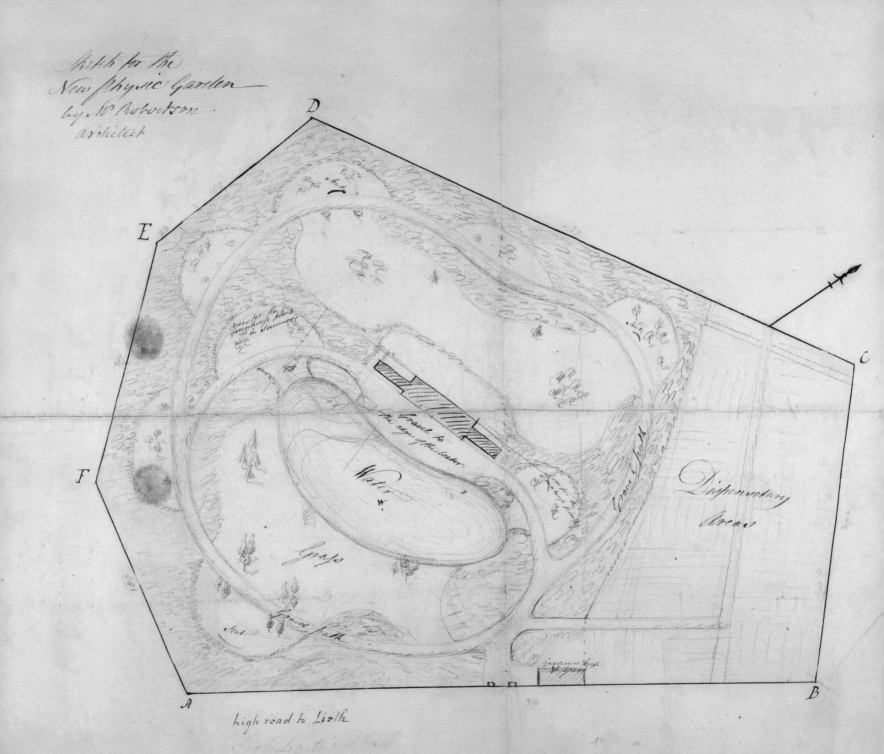

Sketch for the
New Physic Garden—
by Mr Robertson
architect

D

E

C

F

Water

Grass

Dispensatory
Areas

A

high road to Leith

contacts. More will be said of these later, as also of his 1766 trip to England to gain ideas and inspiration for the fledgling enterprise. Not surprisingly, given the scale of the venture, the annual grant for running costs soon proved insufficient to support Hope's untiring efforts at improvement, and in 1774 he judged the time ripe to make a second application to the government for money, through the influence of Margaret Cavendish-Bentinck, Dowager Duchess of Portland, whose son, the 3rd Duke of Portland was then a power in the governing hierarchy. Once again Hope's shrewd timing was successful and he was granted a capital sum of £600 with an addition of £50 a year to the allowance for upkeep. Seven years later the garden was again in financial trouble, and in 1783 we find Hope once more seeking government funding. For a third time he orchestrated a skilful campaign to gain the necessary support to lean on the King and his ministers, once again via the Duchess of Portland. He also sought support from Sir Joseph Banks and the Duchess's chaplain, the Rev John Lightfoot, of whom more anon. Hope was soon able to thank Banks for his efforts, informing him that His Majesty had been pleased to grant the petition for an additional £100. The success of the Royal Botanic Garden Edinburgh is due in no small degree to the good business sense that accompanied Hope's botanical talents and enthusiasm.

Fig. 8. Design for the Leith Walk garden by Robert Robertson, May 1765. Pencil and ink.

DESIGN OF THE GARDEN

For the design of the garden Hope sought suggestions from two gentleman amateurs, Sir James Naesmyth and the advocate Andrew Crosbie. In May 1765 he turned to a professional and paid 'R. Robertson' £2 14s 6d for two designs; these survive, annotated 'Mr Robertson architect', and one of them fairly closely approaches the final layout [fig. 8]. The man responsible is most probably Robert Robertson (1734–1794), confusingly also known as Robinson, a professional architect and garden designer, described by himself as 'late Draughtsman and Executor of the Designs of Lancelot ["Capability"] Brown Esq.', who in 1760 entered a partnership with the nurseryman

and arboriculturist William Boutcher of Comely Garden (near Holyroodhouse). Robertson/Robinson's involvement with Hope's garden has not previously been noted, and much further work is required on what is a historically important question; his work on other Scottish landscape gardens in the 'English' style was treated by Tait (1980).

Two contemporary bird's-eye views of the garden exist; these grisaille wash drawings have recently been discovered to be the work of the great Scottish Neo-Classical artist Jacob More, who was paid a guinea for the pair on 30 December 1771 [figs 9, 71]. This shows not only Hope's interests in the

Fig. 9. Perspective view of the Leith Walk garden. Watercolour by Jacob More, 1771.

visual arts, but the same refinement of taste in choice of artist as with his architect-designers. The drawings, made at an early stage in More's career, just before he left for Italy, are of great interest. More became known for idealised evocations of Classical landscapes but was here asked to depict a landscape with an underlying scientific agenda – somewhat akin to artistic representations of early industrial enterprises, such as iron works or paper mills, or – one that Hope had recently seen – Matthew Boulton's Soho factory. But the practical value of the drawings is that they allow some slight flesh to be put on the bones of the extant garden plans, which have yet to be fully interpreted in conjunction with the descriptive notes in Hope's archive. What is revealed is the revolutionary nature of Hope's creation in terms of its primary role – that of an academic botanic garden. He can have had very few models for such an institution – he knew Paris and owned a plan of Vienna, but until 1766 he had not certainly seen either Chelsea or Oxford (though it is possible that he had plans of these no longer extant in his collection); at this point Cambridge was only just being laid out. But none of these institutions, with their strictly rectilinear plans, with beds of taxonomically arranged plants, provided a model for what is seen in More's views, or in the plan of the finally realised garden made in 1777 [fig. 10]. The rectilinear part is there, on the north-east side, a 'Dispensatory Area' or 'Schola Botanica' where the officinal

plants were grown in Linnaean order (and where they were very squashed, as shown by Hope's meticulous measuring of the placing of plants to the nearest inch) – but it occupies only a small fraction of the site. The rest is an up-to-the-minute landscape garden, where herbaceous plants were grown among trees and shrubs, in compartments between curved and serpentine gravel walks; a formal element is given by the diagonal placing of the conservatories and their relation to a precisely elliptical pond for the growing of aquatic plants (around which were rocks for growing alpines). There are no similarities to Oxford or Cambridge, though there are to the ensemble at Kew – but on a miniature scale. Princess Augusta's botanic and physic garden formed only a small plot at the edge of an extensive landscape, and was more of a private royal affair than a public or academic one. All this must be studied and all that can be given here are a few details and pointers.

Even the shape of the plot is highly unusual – not a right-angle in sight, an irregular hexagon the sides each of different length – none of which was dictated by physical constraints of the site or its ownership (there was plenty room on Hope's 13 acres had he wanted a square). Within the boundary the layout was not only a proto-romantic landscape and a place for teaching, but it also served as a site of commemoration, for the various garden divisions were named after botanists in a carefully thought out way related to the geography and planning

of the plot – from an outer periphery, an intermediate area to the sides and rear of the conservatories and culminating in a Royal Quarter, or Semicircle, in front of them [fig. 11]. Those honoured included not only revered figures of the past, national and international – Sibbald (compartment 1), Ray (16), Tournefort (20), Caesalpinus (23), Hales (25), etc.; contemporary gardeners – Philip Miller (4), Peter Collinson (6), John Fothergill (8) and William Pitcairn (9), but also, demonstrating Hope's humanity and lack of snobbery, his own gardener John Williamson (12). Those commemorated in the royal enclave were clearly of special significance: Hope's patrons Lord Bute (at the centre, 31) and Sir James Naesmyth (29); his teachers Bernard de Jussieu (30) and Charles Alston (32); and George Drummond (many times Lord Provost of Edinburgh, patron of the Infirmary and originator of the New Town, 33). There seems no doubt that Hope was influenced by innovative contemporary garden designs such as The Leasowes, though in an academic context there was a limit to what he could do in the way of eye-catching sculpture, inscriptions or ornament – with two exceptions, monuments to Linnaeus and Williamson, to be discussed later. As an indicator of Hope's interests in romantic landscape gardening his 'Ideas of a Winter Garden' are reproduced in Appendix 2.

Fig. 10. Plan of the Leith Walk garden, dated 6 September 1777. By William Crawfurd. Ink and watercolour.

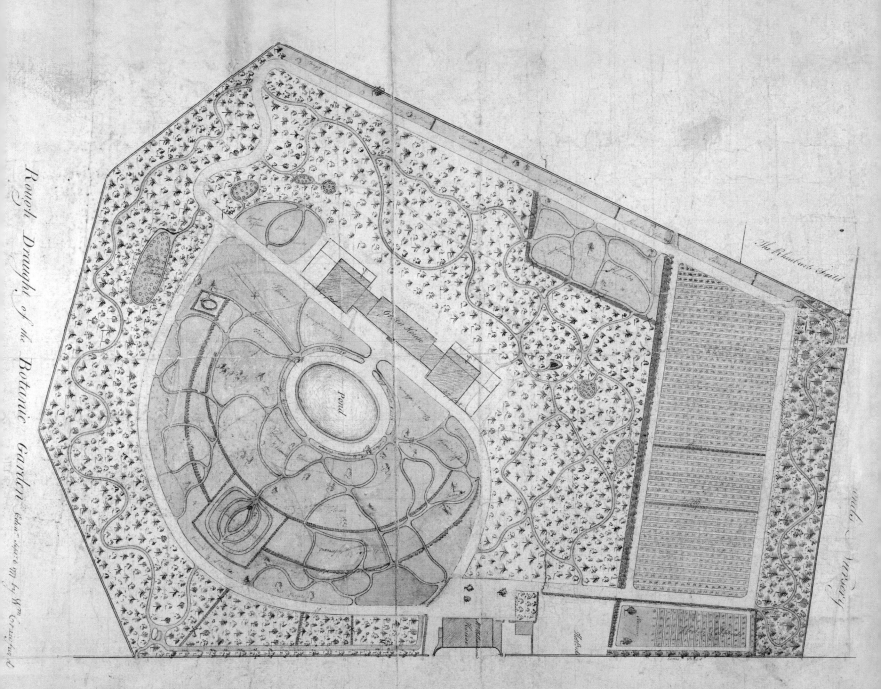

Rough Draught of the Botanic Garden

Road from Edinburgh to Leith

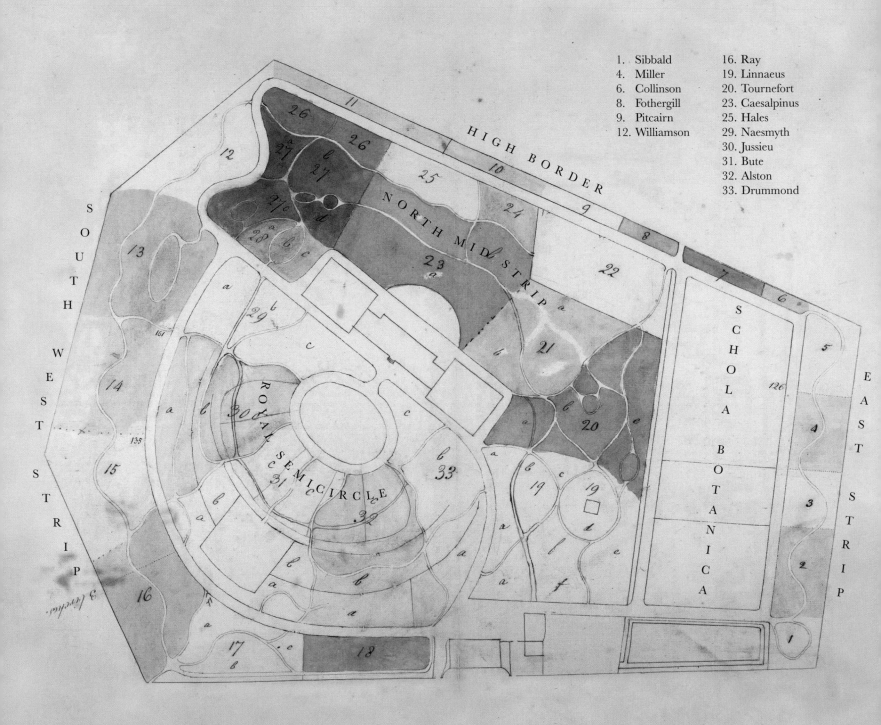

1. Sibbald
4. Miller
6. Collinson
8. Fothergill
9. Pitcairn
12. Williamson
16. Ray
19. Linnaeus
20. Tournefort
23. Caesalpinus
25. Hales
29. Naesmyth
30. Jussieu
31. Bute
32. Alston
33. Drummond

AN ENGLISH EXCURSION

One of Hope's first acts after establishing the garden on the new site was to make, in autumn 1766, a three-week expedition to England to visit gardens, plant collections, nurserymen and museums. The aim was to gain information about garden management and horticultural techniques, and to seek new and exciting plants to add to the Edinburgh stock. Sir James Naesmyth was certainly with Hope for the month in London and on the return journey, and may well have been Hope's companion for the whole excursion – he must surely have helped with introductions to aristocratic garden owners such as the Duke of Northumberland at Alnwick, and it is possible that he sponsored the trip. John Harvey (1981) published the surviving documentation for this important journey in its entirety. On the way south Hope visited the then new 'Walkerian' botanic garden at Cambridge (on a five-acre site in Free School Lane) and sketched its conservative plan. While in London he went to the key nursery gardens (James Lee of the Vineyard, John Gordon's at Mile End) and lesser ones, including John Jeffrey's at Brompton Park, the great private

Fig. 11. Layout of beds at the Leith Walk garden. Anonymous, c. 1780. Ink and watercolour. Captions from Hope's unpublished notes have been superimposed.

gardens of Peter Collinson and Fulham Palace, and the Apothecary's Physic Garden at Chelsea where Philip Miller, author of the famous *Gardener's dictionary*, had been in charge since 1722. This visit by an enthusiastic, and practising, advocate of the Linnaean binomial system of nomenclature may have had important consequences, for it was in the next (8th) edition of his *Dictionary* of 1768 that Miller adopted binary names for the first time. Inevitably Hope also went to Kew, where William Aiton (trained by Miller) had been put in charge by Lord Bute in 1759, with instructions to make part of Princess Augusta's grounds into a physic garden on the pattern of Chelsea. After his visit Hope drew a plan of the Chelsea garden from memory, and made copious notes of plants shown him by Aiton. Near Kew he also visited the great garden at Whitton, laid out in the 1720s by Bute's uncle Archibald Campbell, as Lord Ilay before he became Duke of Argyll. Argyll, as one of his guardians, was a major influence on the young Bute, and after Argyll's death in 1761 Bute moved many of his uncle's rare trees from Whitton to Kew.

Whilst in London Hope sought out progressive scientific company, attending the London Medical Society and dining at the Mitre Tavern with the notable Scottish medical émigrés Sir John Pringle and John Hunter, the anatomist. While in London it is likely that Hope also met the great

American savant Benjamin Franklin. He spent a forenoon (and regretted that it had not been longer) with William Hudson, who had recently published his pioneering *Flora Anglica*, though Hope thought it remarkable that his herbarium was not arranged in a systematic order; he spent another half-day with Stanesby Alchorne who gave Hope specimens from his herbarium of English plants. The British Museum (which had then been open to the public for only seven years) was duly visited and Hope commented on Adam Buddle's British herbarium, and studied the botanical paintings of Nicolas Robert and Maria Sybilla Merian. On the return journey Hope and Naesmyth visited Oxford where they met the Sherardian Professor, Humphrey Sibthorp, saw the Botanic Garden, examined the great libraries and herbaria of Sherard, Morison, Bobart and Dubois, and a collection of Chinese botanical drawings. After a short stop in Stratford on Avon, and at the poet William Shenstone's pioneering landscape garden at The Leasowes, the pair returned home by Birmingham (where they met Matthew Boulton and John Baskerville) and Manchester, making a detour to see the Bridgewater Canal Tunnel and the garden of John Blackburne at Orford Hall near Warrington.

What was not noted by Harvey was that, prior to the English trip, in the autumn of 1765, Hope had undertaken a similar expedition around Scotland, visiting some of the key landscape gardens of the period (and one from an earlier generation): Loudoun Castle, Hamilton Palace and Chatellerault, Barncluith, Inverary, Taymouth, Blair Atholl, Dunkeld and Scone. The network of patrons and garden designers that lie behind the planning of this tour is one of the many aspects of Hope's involvement with progressive garden design that requires further investigation.

STOCKING THE GARDEN

Everywhere Hope went on his English jaunt he made lists of the plants he wanted for Edinburgh. But he had other strategies and means for filling his garden, which was to include both native Scottish plants and exotics, using an extensive network of contacts including former pupils (and the odd diplomat), who were rewarded with gifts of botanical books. The sources included botanic gardens in France, Spain, Italy, Russia and England, and wild material from North America, the Indies West and East, Iberia, Syria (Aleppo), Abyssinia, the South Seas and India. The more important of these sources will be discussed briefly here, starting with the most significant in terms of numbers, which, given the period, were Britain's first Empire – North America and the West Indies.

North America

An important project initiated by the energetic new professor was the founding in 1763–4 of what was probably the first syndicate in Britain for importing exotics: the Society for the Importation of Foreign Seeds and Plants. This scheme was obviously related to Hope's plans for building up what might be called the hortidiversity of the garden, but it also reflected his wider agenda of 'improvement': acclimatising exotic plants of potential economic value and encouraging their transfer between different parts of the expanding British Empire.

Members of the Society paid an annual subscription of two guineas and were entitled to a share at prime cost of all seeds or plants purchased by the Society from collectors in different parts of North America: in addition they received a catalogue of the plants with their botanical names. The first collector the Society dealt with was John Bartram of Philadelphia, recruited by Benjamin Franklin to whom Hope had communicated his proposals. Later there were collectors in other American states and in Canada. For various reasons the Society was dissolved in 1773, but during its existence John Hope was the organiser, dealing with correspondence, payment of collectors, and the sale and distribution of seeds to members. Records of the Society are not extant but the diligence of R.L. Emerson (1982) has

revealed much interesting information about this enterprising venture. It is significant that Hope's first, deeply respectful, letter to the great Linnaeus, in March 1765, was accompanied by a gift of seeds from Quebec sent 'on the instructions of a society set up by certain gentlemen for the importation of exotic seeds into our island'.

A later donor of American seed was Archibald Menzies, who was born in the parish of Weem near Aberfeldy in 1754, and became a gardener in the Edinburgh botanic garden around 1770. The records are not complete, but he was certainly employed, at a weekly rate of 4s 6d, for 33 weeks in 1775, 36 in 1776, 26 in 1777 and two in 1778. Not all the work was within the garden itself, as Menzies collected Scottish specimens for Hope's herbarium and doubtless living plants for Leith Walk; the 1778 payment may be for the trip he made that year to collect Scottish plants for the London gardens of Drs John Fothergill and William Pitcairn. His talents were greatly appreciated by Hope, whose botanical lectures he attended in 1773 and 1774, and it seems likely that Menzies's attendance at other medical faculty lectures between 1771 and 1774 and from 1777 to 1779 was subsidised by Hope, though he did not graduate. The first part of Menzies's career was as a naval surgeon, which allowed him to introduce some notable plants to British horticulture, most famously the monkey puzzle (*Araucaria araucana*) from Chile.

His first voyage, in 1782, was to North America, where from 1783 to 1786 Menzies was assigned to HMS *Assistance*, based in Halifax, Nova Scotia. From Halifax he sent seeds for the Leith Walk garden (and to Banks for Kew) that he had collected in the Caribbean, New York and Canada. In Nova Scotia he was thrilled with the rich temperate flora encountered – conifers, Ericaceae and cryptogams – and it is worth quoting the heartfelt gratitude that Menzies wrote to his benefactor in May 1784:

> In this situation the tears trickled down my cheeks in gratitude to you, Sir, who first taught me to enjoy these pleasures which Providence has so conspicuously placed before my eyes, accept of them as the only mark a grateful heart can at present offer. I shall wholly devote my vacant hours to Natural History while I remain on this station and I have no doubt but that I shall be able to send you another parcel of seeds and specimens early in the Autumn – if in this country I can in any other respect serve you, you have a just title to command me, and I shall ever think it my duty to obey (quoted in McCarthy 2008).

Hope thanked Menzies in 1785 with copies of the latest (12th) edition of Linnaeus's *Systema naturae* and his own *Catalogue* of trees and shrubs in the Edinburgh garden. After this first trip Menzies was to circumnavigate the globe twice, first on a fur-trading voyage that left Britain in September 1786, but Hope had died before

Menzies even reached North America for the second time. The third voyage was the famous one of 1791–5 with Vancouver on *Discovery*. On these voyages Menzies made extensive collections of plants and seeds and when he died, in 1842, he bequeathed his herbarium of grasses, sedges and cryptogams to RBGE, including some algae from the Firth of Forth [fig. 23] and grasses and sedges from the Leith Walk garden collected when he worked there 70 years earlier [fig. 12].

The West Indies

In Hope's period the West Indies included some of Britain's most important colonies: the source of immense wealth, not least to Scotland. Whatever today's moral repugnance at the methods used to cultivate and extract this wealth (and contemporary practical difficulties arising from fighting with the French), the islands of Antigua, St Kitts, Jamaica and Barbados were a major source of students for the Edinburgh medical school,

Fig 12. Specimens collected from Leith Walk garden by Archibald Menzies in 1777 and 1779. *Cyperus compressus* (left), and *Eragrostis flexuosa* (now *E. cilianensis*) (right). These are both widespread, weedy annuals that could have been sent to Hope from almost anywhere in the tropics, but possibly came from India.

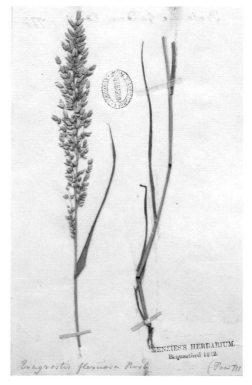

and provided Hope with a valuable network of medics and former pupils to supply him with exotic seeds.

In 1773 Hope wrote to the Governor of Jamaica, Sir William Trelawney, requesting seed 'of every Tree, Shrub or beautiful Plant, particularly such as are used in food, medicine or any œconomical purpose', which would be 'acceptable & useful in this Garden'. By this time Hope was already in contact with William Wright, an Edinburgh-trained surgeon who had been in Jamaica since 1765. Wright returned to Edinburgh for a medical refresher course in 1778, when he met and talked with Hope though he did not, as initially intended, attend Hope's botanical lectures. While in Jamaica Wright assembled several remarkable bound sets of dried plant specimens (*horti sicci*) copiously annotated with ethnobotanical and medical information – Hope had a two-volume version from which he made numerous memoranda for incorporation into his lectures. There is no record that Wright sent seeds to Edinburgh, but around this time Matthew Wallen, another Jamaican botanist, did and was sent a gold medal for his trouble.

The richness of the Caribbean as a source of tropical plants led Hope to recommend two of his best students to go to Jamaica. The first of these was Thomas Clerk (later consistently spelt Clarke), who had taken the Botany course in 1772, 1773 and 1774, and who made microscopic studies of pollen for Hope. Hope must have rated Clarke highly as he sent him with an introduction to Linnaeus, and then to Banks – the latter meeting resulting in his appointment in 1777 as the first government botanist in Jamaica.

John Lindsay was an even more distinguished pupil, starting out as a 4/6-a-week 'gardener' at Leith Walk from 1773, and still being paid at the garden in 1779 (his pay having risen to six shillings in 1778). According to Edinburgh University records Lindsay attended Botany lectures in 1776 (with Maths and Natural Philosophy) and 1777 (with Anatomy and Surgery), and lectures in Materia Medica and Practice of Medicine in 1779; that Hope did not mention him in his own class lists (even under the 'gratis' category), and that Lindsay referred to Hope as his 'patron', suggests that, as with Menzies, Hope probably subsidised his medical training, though, as also with Menzies, Lindsay was never to graduate. Lindsay's work for Hope was far more than horticultural – he drew many teaching drawings and also botanical illustrations including *Buddleja globosa*; it has also recently emerged that it was he who performed the highly original light/gravity experiments to be discussed later. Taking precious seed with him, including the 'moving plant' (now *Codariocalyx motorius*), Lindsay went to assist Clarke and was in Jamaica by 1782. The quantities of seed he sent Hope were (as with several other donors) repaid with gifts of books; Lindsay also sent back a drawing and description of a new *Cinchona* (now *Exostema brachycarpum*) [fig. 13]. Lindsay's most important research in Jamaica was his pioneering microscopic observations of the germination of fern spores, and his experimental work on the sensitive plant (*Mimosa pudica*). He communicated the fern observations to his revered teacher, who, after Lindsay had satisfied various queries, agreed to present them to the Royal Society of Edinburgh. Hope's death intervened so Lindsay sent his paper to Banks who read it to the Linnean Society, and it was published in their *Transactions* in 1794. The work on *Mimosa* was also sent to Banks, but for some unknown reason was never published, though the manuscript is still at the Royal Society.

Hope's final important West Indian contact was Alexander Anderson, who took over the running of the twenty-year-old botanic garden on St Vincent in 1785. Anderson is said to have studied medicine in Edinburgh, but does not appear on any of Hope's class lists, and their connection was probably through Banks. In 1786 Anderson sent Hope a catalogue of the St Vincent garden and seeds, and, had Hope lived, would doubtless have become a major correspondent, for he was to play an interesting role in pioneering forest conservation efforts in the Caribbean (Grove 1995).

India

Fortunately for the British economy, as the American empire waned another was waxing in the East. About thirty of the students who attended Hope's lectures between 1761 and 1786 took advantage of this and joined the East India Company as assistant surgeons, where (amongst much else) they made very significant contributions to the study of the Indian flora, running botanic gardens, and developing the economic potential of Indian plants.

In the present context, that of stocking Hope's garden, it was India that supplied him with one of his greatest marvels, the 'moving plant' [fig. 14]. This came from Bengal, sent by James Kerr as the *burrum chundalli*, and it was under this exotic name that the plant was recorded in one of the few contemporary accounts of the garden – in the informative *History of Edinburgh* by the advocate and antiquarian writer Hugo Arnot (EP21) published in 1779. Scientifically Hope considered this legume to be a *Hedysarum*, but it is now placed in a closely related genus as *Codariocalyx motorius*. One of Hope's

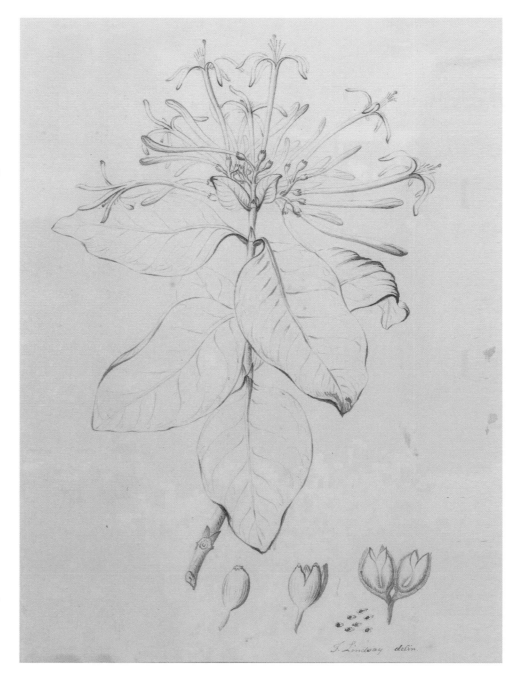

Fig. 13. *Cinchona brachycarpa*. Drawing by John Lindsay, who discovered the tree in the parish of Westmoreland, Jamaica, in 1784. Brown ink.

This drawing was sent to Hope, with a Latin description and herbarium specimen. Lindsay's description was not published until 1794, by which time this drawing had been mislaid, and Olof Swartz had pre-empted the name in 1788; it is now placed in a related genus as *Exostema brachycarpum*.

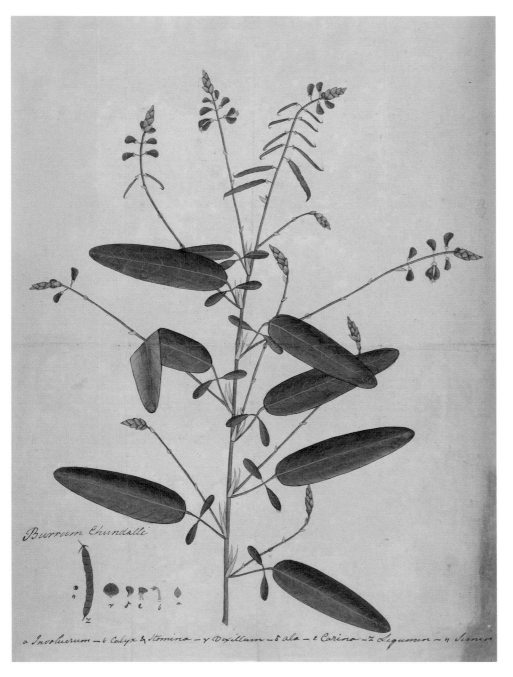

Burrum Chundalli

a Involucrum – b Calyx & Stomina – γ Vixillum – δ ala – ε Carina – z Legumen – η Semen

interests was in similarities and differences between plants and animals, and hence the phenomenon of movement in plants. This plant demonstrates two distinct sorts: at night the large central leaflet declines in a 'sleep' movement, but, more curiously, the two small basal leaflets make spontaneous jerks during the day (for which reason it is also known as the 'telegraph plant'). This made it an object of superstition in Bengal, though the 'Natives' may have been having a joke at Kerr's expense when they told him that on a Saturday they would 'cut off [the] two [basal] lobes the instant they approach together, & beat them up with the Tongue of an Owl: with this composition the Lover touches his favourite Mistress, to make her comply with his wishes!' When he attended Hope's lectures in 1774 Kerr was already an experienced surgeon, both in India and on the high seas; he must have been talented, as Hope asked him to investigate pollen microscopically with Thomas Clerk.

With the seeds and a written description, Kerr also sent back to Edinburgh a drawing of the *burrum chundalli* by an Indian artist [fig. 14]. This use of indigenous artistic skills to supplement specimens and written descriptions was developed on a spectacular

Fig. 14. 'Hedysarum movens', *burrum chundalli*. Drawing by an anonymous Bengali artist, c. 1775. Bodycolour heightened with gum arabic.

Hope's sensational 'moving plant' sent from Bengal by James Kerr; it is now known as *Codariocalyx motorius*.

scale by another pupil who went to India and made a fortune by metaphorically 'shaking the pagoda tree'. This was William Roxburgh from Ayrshire, who attended Hope's lectures in 1772 [fig. 15]. Roxburgh started off in Madras, and became known as the 'Father of Indian Botany' for his pioneering efforts to document the Indian flora, using the Linnaean system, but, unusually, with plant descriptions written in English. The result, his *Flora Indica*, was published in its entirety only in 1832, 17 years after its author's death (Robinson 2008). Of the 2500 drawings commissioned to accompany this work, first in southern India, but continuing after Roxburgh became Superintendent of the Calcutta Botanic Garden in 1793, only 300 were published in his lifetime – by Sir Joseph Banks, in lavish imperial folio format, as the *Plants of the coast of Coromandel* [frontispiece]. Roxburgh sent Hope not only notes on useful plants including sandalwood and the marking nut tree, but seeds of plants from Madras to be grown in the Leith Walk conservatories. As with Lindsay in Jamaica, these were repaid with medical and botanical books unobtainable in remote colonies (in Roxburgh's case with Sir John Hill's edition of the first volume of *Hortus Malabaricus*).

One of the most innovative areas of Hope's teaching related to natural classification, as opposed to the artificial sexual system of Linnaeus. This found a willing convert in the person of Francis Buchanan, who attended Hope's class in 1781 and left the best account of his natural system in any of the student lecture notes. Buchanan went to India, where he made pioneering studies not only of botany, but in the 'statistical' tradition of land use, customs, religion, minerals, agriculture and archaeology, of several fascinating and under-explored areas including Burma, Nepal and Mysore. Unfortunately Buchanan gave the specimens and drawings from these expeditions to others (the Burmese collection to Banks and the Nepalese and Mysore ones to his Edinburgh contemporary J.E. Smith), who failed to do much with them. Buchanan's own manuscript Floras of Mysore and Nepal (with Latin plant descriptions) have never been published, but are of great interest for their early use of A.L. Jussieu's natural classification system. Like Roxburgh, Buchanan also made extensive use of Indian artists – not only for plants, but to draw birds and mammals whilst he was superintendent of the Barrackpore Menagerie for the Governor-General, the Marquess Wellesley, and also the fish of the Gangetic river system, on which he published an important account, resulting from his final great statistical survey of Bengal.

Edinburgh students in India made major contributions not only to what would now be called the cataloguing of biodiversity. As already noted, a major interest of Hope and his contemporaries was the applied (including medicinal) use of natural

Fig. 15. William Roxburgh (1751–1815). Engraving by C. Warren (photographically enlarged reproduction from *Annals of Royal Botanic Garden Calcutta* vol. 5, 1895).

products. Two students made particularly important contributions to this latter subject in India – John Fleming (who also ran the Calcutta Botanic Garden on two occasions) in 1810 compiled the first English-language catalogue of Indian drugs, and in Madras, in 1813, Whitelaw Ainslie from Duns published the important *Materia medica of Hindoostan*.

Brucea antidysenterica.

Africa

One of the most surprising recent findings about plants grown at Leith Walk is that there were at least six species from Abyssinia (Ethiopia) donated by the great traveller James Bruce of Kinnaird (EP15). In fact these were specially mentioned by Hugo Arnot (1779), but this had since been forgotten. Hope's private papers, and notes from his lectures, show not only that Hope was proud of these rare additions to his garden (for which he gave Bruce a gold medal), but that he was a friend of Bruce and treated his second wife (by correspondence) during her terminal illness in 1785. Bruce's travels to the source of the Blue Nile between 1768 and 1773 became notorious when his account eventually appeared in 1790, and doubts were cast by Dr Johnson and others on the veracity of some of its more sensational contents – most of which have subsequently proved to be true. Bruce took an Italian artist, Luigi Balugani, with him, who drew plants and animals, of which the former are now in the Yale Centre for British Art (Hulton *et al.* 1991). Among this collection is a drawing of the cereal 'teff' (*Eragrostis tef*) made at Leith Walk by Andrew Fyfe in 1775. A search in the RBGE herbarium has

Fig. 16. *Brucea antidysenterica*, 'wooginoos'. By Frederick Polydore Nodder, 1777. Watercolour, bodycolour with pen and black and brown ink, over graphite.
Yale Centre for British Art, Paul Mellon Collection.
One of James Bruce's Abyssinian plants grown at Leith Walk.

uncovered a specimen of this grass collected in the garden by Archibald Menzies in 1777. The other African treasures included the exotic-sounding 'wooginoos', subsequently named after its finder as *Brucea antidysenterica* [fig. 16], a plant that did what its Latin name claimed – taken in camel's milk it cured the traveller of dysentery.

The South Seas

Another group of plants mentioned by Hope in lectures and unpublished notes were those collected on the voyages of Captain Cook. These included material from the *Endeavour* voyage (1768–71) made by Joseph Banks and the Linnaean pupil Daniel Solander. Hope was given some of their herbarium specimens, and at least three living plants: 'a new guava' (probably *Psidium guajava*), New Zealand tea (*Leptospermum scoparium*) and 'Mr Banks' new tree with Equisetum leaves' (*Casuarina equisetifolia*). Solander, by now Banks's librarian, also provided Hope with botanical information, including names for Bruce's plants, and descriptions of six North American trees and shrubs that Hope published for the first time (probably unintentionally) in the second edition of his garden *Catalogue* ([Hope] 1778). Unreported until now is that Hope also had plants collected by William Anderson, who was surgeon on the second (*Resolution* and *Adventure*, 1772–5) and third (*Resolution* and *Discovery*, 1776–80) voyages, though Anderson,

like Cook, did not return from the latter. For these gifts Anderson was another recipient of a gold medal.

The only species that Hope described new to science was a now popular garden shrub, with flowers in marigold-orange balls, *Buddleja globosa*, a native of Chile and Argentina [fig. 17], but there have always been several puzzling aspects to Hope's skeletal published description of the plant. First the place of publication – the journal of the Royal Dutch Society of Sciences and Humanities; second, the unknown source of Hope's plant, assumed to have been the London nursery credited in *Hortus Kewensis* with its introduction in 1774. The place of publication makes a tenuous link with the Dutch banking branch of the Hope family – as one of its male members presented copies of this same journal to the Royal Society of London; that Dr John Hope retained links with them is shown by his medical treatment of a female member of the family on her visit to Edinburgh in 1785. It has now been found (in notes from Hope's lectures) that the *Buddleja* was one of the plants given to Hope by William Anderson, and must therefore have come from the second Cook voyage – and Hope wrote to ask Banks if he had seen it on the first voyage. But this merely leads to a further mystery, as on Cook's second voyage the only parts of Chile visited (Christmas 1774) were the

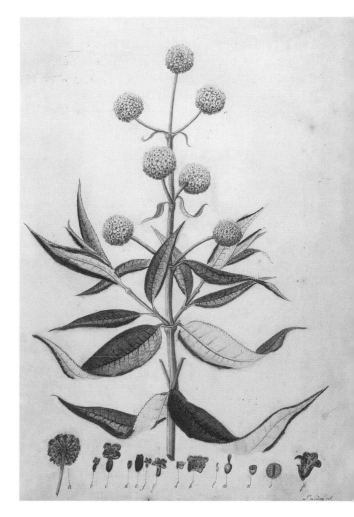

Fig. 17. *Buddleja globosa*. Drawing by John Lindsay, c. 1780. Ink wash.
A manuscript description, probably in Lindsay's hand, explaining the floral dissections accompanies this drawing. This is not the drawing engraved to accompany Hope's published description of the plant in 1782.

islands to the south of Tierra del Fuego, which is about 800 miles beyond the southern limit of the plant's present distribution. No part of Chile was visited on the third voyage, even if Anderson did manage to send Hope anything from it, and Argentina was visited on neither voyage.

Two medicinal plants: rhubarb and asafoetida

Throughout his life Hope retained an interest in medicinal plants, and the introduction and cultivation of two exotic species, then valued items on the drug-list, particularly attracted his attention and became subjects of two of his meagre output of scientific publications (see Appendix 1 for a listing). These were the medicinal rhubarb, *Rheum palmatum*, and an umbelliferous plant later described as *Ferula persica*, source of the aromatic gum 'asafoetida'.

In 1763 Hope was given seed of the rhubarb, *Rheum palmatum*, by Dr James Mounsey, who had returned to Scotland after twenty-five years' medical service in Russia, bringing a box of seed collected from a plant in the St Petersburg physic garden (Appleby 1983). Within a couple of years Hope had *Rheum* plants flowering at Leith Walk and began distributing seeds to friends and acquaintants, including the Earls of Hopetoun and Atholl, Richard Pulteney, Bernard de Jussieu and

John Ellis. In extending the cultivation of rhubarb Hope had the enthusiastic cooperation of his friend the physician Sir Alexander Dick, who had also been given seeds by Mounsey and grew the plants at Prestonfield House near Edinburgh. The roots of home-grown rhubarb were shown by Hope to possess exactly the same medicinal effects and potency as the expensive product known as Turkey rhubarb, which had been imported into Europe from Central Asia since Roman times. In a commercial test Hope recorded that a profit of £12 8s 8d was made by the sale in London in 1778 and 1779 of rhubarb from a plot sown in 1770, the costs of production and transport being £35 4s 4d. By 1782 Hope was able to report that 'now scarcely a garden in Scotland is without a rhubarb plant in it, the consumption of the foreign rhubarb is considerably less, and annually a small quantity is sent to London'. Only local rhubarb was used in the Edinburgh Royal Infirmary, and Hope had about 3000 plants growing in a plot of land just outside the Leith Walk garden. This commercial development of medicinal rhubarb was short-lived, partly as hinted at by Andrew Duncan, because of resistance from trading interests in foreign 'Turkish rhubarb', and partly owing to a decline in its medical use (see also Turner 1938). Hope's brief account of the work with *Rheum palmatum* was published by Banks

in the *Philosophical transactions* (Hope 1766). Hope commissioned a watercolour of this special plant by the noted Edinburgh artist William Delacour [fig. 18], who was the first drawing master at the Trustees Academy; the drawing was engraved by Andrew Bell, and Hope proudly sent copies of the print to Linnaeus, Pulteney and others.

In 1777 Hope received two roots of 'asa fœtida' from Dr Matthew Guthrie, a former student, and another of the series of Scottish physicians who, at various times during the eighteenth century, chose to practise their craft in Russia. The roots were sent in return for seed presented to Guthrie via Hope's kinsman Lord Hope (later 3rd Earl of Hopetoun. EP23). The cultivation of *Ferula* in Scotland was a success, producing both gum of high medicinal quality and fertile seed. In early spring 1783 Hope distributed the latter to intimates in Edinburgh including Sir Alexander Dick and Dr William Cullen, and to people of influence further afield such as Sir Joseph Banks, Richard Pulteney, the Earl of Bute, the Duchesses of Portland and Buccleuch, and William Forsyth at the Chelsea Physic Garden. Despite this, it appears that commercial production of asafoetida in Britain was not as successful as rhubarb, but a brief, illustrated account of his work on the plant was published (Hope 1785).

HOPE'S GARDENERS

John Hope mixed with the intellectual élite of Edinburgh and was related to several Scottish aristocratic families, whose members he advised medically. Yet, although this was an age when distinctions of rank were finely drawn and strictly maintained, this clearly had little effect on Hope's interactions with those of humbler background, whether promising students, employees or patients. This ability to get on with, and get the most out of, artisans is most clearly shown in the relationships he had with his gardeners.

First were the head (principal) gardeners, on whom Hope greatly relied – three of whom, in succession, would serve him well. The first was John Williamson, who, at his death in 1780 had been in post for 20 years, so must previously have worked in the Abbey and Town gardens and supervised the great move. His ability, industry and loyalty were essential in the laying out of the new garden and its subsequent development. He was the first inhabitant of the house on Leith Walk, and though this building (as will be seen) served other functions, it is at least in part a tribute to Williamson that Hope took such care over

Fig. 18. *Rheum palmatum*. By William Delacour, c. 1765. Watercolour.

An engraving of this drawing was made by Andrew Bell, published in the *Philosophical transactions* vol. 55 (1766), and in a slightly different version in the first edition of the *Encyclopaedia Britannica* (1771).

its design and planning. Williamson helped Hope with his experimental work, setting up and monitoring some of the long-running experiments on transpiration and plant growth. One manuscript recording such experiments has survived at RBGE, some of which is in Williamson's hand (spelling was not his strong point). Williamson also undertook hybridisation experiments between the oriental and the opium poppies, though Hope neglected to mention his name when reporting these to the younger Linnaeus. It was probably also the head gardener's job to sort and lay out the specimens and drawings to be shown in lectures and refile them after use, about which Hope was very particular. The head gardener's annual salary was £25, but pocket money was to be made by showing favoured guests around the garden, selling publications to students, and possibly also propagating and selling some plants on the side.

Williamson, however, had another official job that almost doubled his garden salary; he was also a customs officer, or 'land carriage-waiter', at Leith, which must have been useful in the constant toing and froing to Edinburgh's port in connection with building materials, books, seeds and plants brought in for Hope's garden by ship from London (a trans-shipment in the case of foreign material). This second job was to have tragic consequences on 23 September 1780. In Princes Street at 4 o'clock in the morning Williamson chased

Fig. 19. Monument to John Williamson (d. 1780). Possibly designed by James Craig, 1781. Transferred to RBGE from the Leith Walk garden in 1823.

a group of armed smugglers and was 'beat and wounded in so cruel a manner, that he died a few hours thereafter, leaving behind him a wife and three helpless children, one of whom blind from his infancy'. Of these offspring, Williamson's daughter Agnes ('Nancy') drew some of the teaching drawings for her late father's employer and married another of his draftsmen, Andrew Fyfe; the sighted son Samuel trained medically (aided by Hope) and became an East India Company surgeon.

Hope was much affected by Williamson's death and expressed his gratitude posthumously in a handsome stone mural tablet to his memory (which makes no

allusion to his sticky end) [fig. 19]. This cost £2 17s 8d; its designer is unrecorded, but is likely to have been James Craig. In the RBGE archives are two handsome designs for amending the Leith Walk frontage of the garden by aggrandisement of the two humble doors into the garden depicted in More's 1771 drawings; presumably related to these is a design for a strip of garden with a central circular bed as a buffer between the garden and the street – these three drawings have been attributed to James Craig (Cruft 1995). What is of interest is that one of the facade designs shows inscribed tablets of exactly the same form as the Williamson monument placed between the doors and their pediments [fig. 20]. Cruft not unreasonably speculated that these might originally have been intended to receive lines from *The Seasons*, by Craig's uncle, the poet James Thomson (of whom Craig was inordinately proud). It seems likely, however, that not only was this design realised, but that the inscription on one

of the tablets turned into Williamson's epitaph – proudly displayed on the outside of the garden, celebrating the productive collaboration between botanical professor and gardener.

In the Hope collection are two finely drawn designs for mural monuments. The more elaborate (reproduced in Cruft 1995) is signed by James Craig, but the simpler one is almost certainly also his. The latter, a portrait-format tablet with a pediment supported on a pair of consoles, may well be an alternative design for the Williamson monument – the upper part of

the sheet has been removed, and perhaps bore the executed version that was sent to the mason's yard. It is possible that the signed design – a magnificent nine-foot-high structure surmounted by a two-dimensional obelisk – was also a proposal for Williamson's monument, though it is perhaps more likely to be for a monument to Linnaeus (see later), for whom the grandeur would be more appropriate.

After Williamson's death John Bell, who was already working in the garden, acted as head gardener, but stayed only about a year (1781–2) – nothing is

Fig. 20. Design for improvement of the Leith Walk frontage of the gardener's house and garden doors, probably by James Craig, c. 1780. Ink.

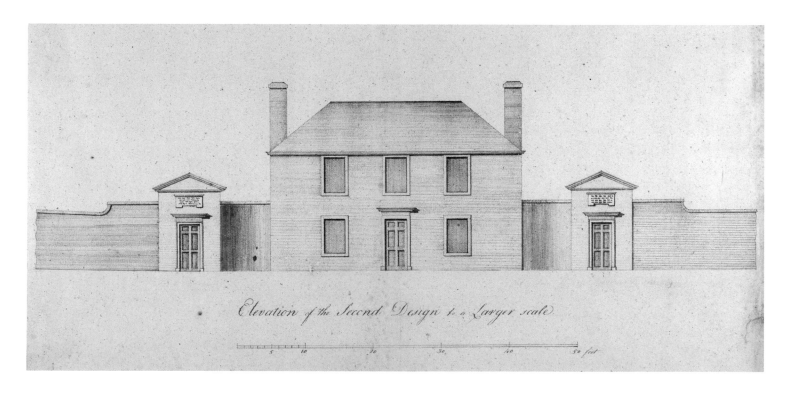

Elevation of the Second Design to a Larger scale

known of Bell, other than that he too must have been multi-talented, as his name appears as artist on some of the teaching drawings [fig. 55]. Bell was succeeded in 1782 by Malcolm McCoig, who lasted for the rest of Hope's tenure, dying in 1789. It is probably McCoig who appears in the Kay caricature [fig. 1], in conversation with his master. There are several letters and manuscripts in McCoig's bold hand among Hope's papers, which show that he was every bit as painstaking in his care of the garden and its plants as Williamson, but somewhat more literate.

Between the head gardener and the labourers who did more manual tasks, augmented in number for big projects, such as digging a major new drain that greatly exercised Hope in 1785, was a second tier of gardeners. This designation, however, is scarcely adequate as these gardeners doubled as plant collectors, research assistants and draftsmen. They were employed on a weekly basis, the basic rate being four shillings and sixpence; they tended to work for long blocks of time, but which allowed them to engage in other activities, in particular the taking of university courses, which Hope encouraged (and probably supported) some of them to do. The most important of these gardener-assistants were Archibald Menzies, John Lindsay, James Robertson and Andrew Fyfe.

Of these, the first three are discussed under the headings of the parts of the world, their careers launched, from which they would send Hope seeds – North America, Jamaica and Scotland respectively. But Andrew Fyfe (or Fife) (1754–1824) falls into a different category.

Fyfe appears as a gardener from 1772 to 1775, and that he performed at least some horticultural roles is suggested by a note that he collected plants during a visit of Hope to London in 1773. Like Menzies and Lindsay, Fyfe was paid weekly at 4s 6d, the periods of work varying from six weeks in 1775 to ten months in 1773. But Hope used his considerable artistic skills to make teaching drawings 'to be seen at any distance in the [lecture] room'. These drawings included the spectacularly rococo 'three inarched trees' in red chalk [fig. 21], ink wash diagrams, and fine botanical illustrations such as ones of asafoetida sent to Banks, and *Eragrostis tef* sent to James Bruce. Fyfe studied drawing at the Trustees Academy, which had been set up in Edinburgh in 1760 by the Board of Trustees for Fisheries, Manufactures and Improvements in Scotland, where he must have been taught by Alexander Runciman. In 1775 Fyfe won the Academy's annual prize medal. As with Lindsay, Fyfe does not appear on any of Hope's own class lists, but Edinburgh University records

suggest that he took the Botany class in 1776. His attendance at other medical lectures (Anatomy and Surgery in 1775, 1779, 1780, 1781; Chemistry and Materia Medica in 1775; Theory and Practice of Medicine in 1776) is likely to have been subsidised by Hope. In 1777 he became 'principal Janitor and Macer [mace-bearer]' at the University, and Dissector in Anatomy – posts that he held for forty years, the latter under the Monros *secundus* and *tertius*. Fyfe also lectured on anatomy, but is better known for two text books (in several editions) illustrated with engravings from a variety of sources, some after his own fine drawings (Rock 2000). In 1787 Fyfe married Agnes Williamson, a romance that must have blossomed over the teaching drawings they made for Hope. It would be fascinating to step back into the gardener's house in the 1770s and observe this network of relationships at the intersection of academia, art and horticulture.

Fig. 21. Three inarched trees. By Andrew Fyfe. Red chalk.

Highly elaborated from an engraving by Simon Gribelin, Plate 11 of Hales's *Vegetable staticks* (1727). The experiment shown was repeated at the Leith Walk garden, where three willow trees were grafted together and allowed to grow; the soil was then dug from beneath the central one, which continued to grow (suspended), due to lateral movement of sap from the outer trees. The drawing is annotated 'this is shewen at the lecture on the motion of the sap'.

BOTANIC COTTAGE

A major part of the information on Hope and his Leith Walk garden discovered since the first edition of this book has arisen from the project centred on what by the mid-nineteenth century had come to be known as Botanic Cottage. This little building survived the removal of the garden to Inverleith under Robert Graham in the early 1820s. Despite all Hope's efforts his garden turned out to have a lifespan of a mere sixty years, its abandonment due to fumes from an iron foundry on the opposite side of Leith Walk, projected building activities on the surrounding land and anticipated smoke pollution; a further problem was that access to the garden had been greatly impeded by the raising of the level of Leith Walk prior to 1815.

What was left of the cottage in 2008 was but a battered, unloved fragment of its original self – all that was visible from Leith Walk was its upper storey, entered by a concrete bridge through what was once an upstairs window; in 1912 its south-west gable had been demolished to build a tenement that loomed boorishly over its once charming neighbour. With the threat of demolition in 2007 a team of enthusiastic local residents led by Eileen Dickie of the Friends of Hopetoun Crescent Garden got together to try to find a new use and purpose for the building. The Friends had originally

been set up to preserve the sole remaining fragment of Hope's garden, an arc that once formed part of its north-west side (a section of the High Border) retained as a green prospect for an 1820s building development, originally called Hope Crescent. The group was fortunate to secure the services of conservation architect James Simpson, who made the bold suggestion that if the building could not be saved, it could be moved, stone by stone, and accurately recorded during the process. The Friends obtained a Heritage Lottery Fund grant, which enabled archival research on the building to be undertaken by Jane Corrie and Joe Rock; what follows, with their kind permission, is largely a summary of their work. The other major source used is the report of the archaeological survey of the building undertaken by the Glasgow University Archaeological Research Division in 2008 (Smith & Francoz 2009).

When built in 1764, the cottage was a handsome Georgian doll's house of three bays, with a piend (*Anglice* 'hipped') roof, and a chimney stack on each gable. The symmetrical street facade had a central door, a window at either side and three at first-floor level – it was linked to the garden wall on either side by a small quadrant. Adjacent to each quadrant lay a door, the left-hand one opening into the garden, the right-hand into the gardeners' yard. The original rear

elevation is less certain, as it was altered on various later occasions. Like the front, it originally had five windows and a door, but when first built the latter was, oddly, asymmetrically placed (to the left), a position almost immediately changed to the centre. Unlike the front facade the upper windows were not identical, the central one being narrower and longer (floor to ceiling); Tom Addyman has suggested that it might have had a balcony from which to view the garden – but it could simply have been longer to light the staircase. The original staircase was internal and wooden, but was almost immediately rebuilt in stone, leading from the gardener's quarters below to the upper floor – the change in position of the back door was connected with this, but the details remain obscure. A great change took place after Hope's time, in 1802, when an external stair tower was added to the centre of the rear facade, doubtless to increase space inside the cottage.

Substantial records survive of payments to the tradesmen who built, and very actively maintained, both the cottage and greenhouses (the latter are not discussed here, and much work remains to be undertaken on them), but even so puzzles and uncertainties remain, and there is an unfortunate lack of contemporary visual evidence. All that can be given here is the merest summary, concentrating on aspects relating to the functions that the building performed for Hope.

As noted above, it seems fairly certain that John Adam, or a member of his office, was responsible for the design of the cottage. The main building work was undertaken in 1764 by the mason James McPherson; the construction was of random rubble, originally covered with a flush, lime-mortar finish to contrast with the dressed Craigleith stone used for the quoins and the window- and door-surrounds; the roof was of Easdale slate, for which James Ramsay was paid £20 13s 4d on 15 November 1764. In 1765 work on the interior was in progress – windows, flooring and plastering – and in March, John Young, wright, made a wooden staircase that was rebuilt in stone by McPherson a month later; lead came from the Hopetoun family mines around Leadhills in Dumfries-shire. Of greatest interest in terms of Hope and his teaching is the discovery that the upper floor consisted of a single Great Room, about 32 feet long and 18 wide; it had a wooden stage, from which Hope delivered the first, and at least some of the second, part of his lecture course before heading out into the garden and greenhouse for the third. A small section of this room may have been partitioned off with a wooden screen to form a small room for Hope, as in 1768 there is mention of making book shelves

for the 'Professor's room'. In 1785 Young made alterations, including 'openings' in a (?this) partition, and 'making and hanging blinds' on the stage – perhaps the supports on which Hope's large teaching charts were displayed, attached by means of cloth loops [fig. 56].

The only decorative element of the original scheme may have been a railing around the head of the staircase, which, somewhat whimsically, was in the Chinese style, and later painted green. The ceiling of the Great Room was 'coombed', the lower part sloping inwards from all four walls, to give greater height and a sense of space – it must have been quite crowded with classes of up to 102 students. The earliest record of the decoration of the lecture room is from 29 September 1767, when the ceiling was painted white and the walls 'yellow syze'. This is of great interest as the invoice is to 'Alexander Runciman painter'. Runciman is a noted Neo-Classical artist, though in June of that year had embarked on his travels to Italy, so the work must have been undertaken by his partner Dugal Maclaurin, who signed for the work. Maclaurin was the son of a surgeon, and Hope had stood cautioner when he was apprenticed in 1759. The room was repainted yellow in 1785 by Thomas Peacock. For lectures the students sat on long wooden forms, but Hope's purchase of eight beech chairs

with stuffed seats from Alexander Peter in 1766 suggests that the posteriors of gentlemen attenders of his classes, such as the advocate Andrew Crosbie and the banker Sir William Forbes (EP16), were more comfortably accommodated.

Details of the downstairs rooms are sparser, though that on the right-hand side was referred to by Hope as the parlour, and to this, in 1780, was added a new room built out into the yard, entered through a door knocked through the north gable. The usage of the space to the left of the central corridor is uncertain, but must have comprised the kitchen and sleeping rooms for the gardener and his family (and it is possible that this was, at least later, the location of the Professor's room).

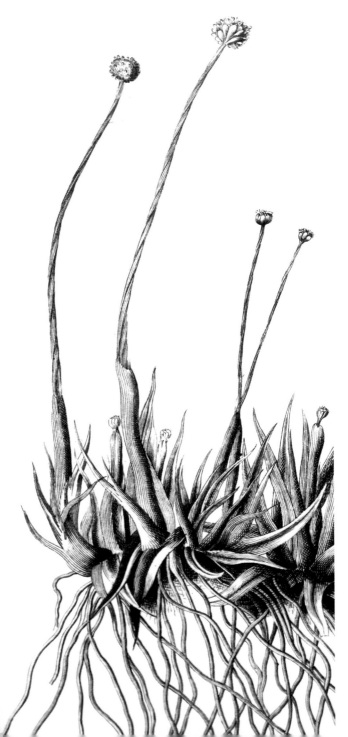

Hope and the plants of Scotland

The most immediate source of living plants for the Leith Walk garden was Scotland itself, and the garden's early financial accounts record quite substantial sums paid for making such acquisitions – those of Archibald Menzies have already been mentioned, but more significant were those of James Robertson, to be discussed later.

Hope's interests in the Scottish flora, however, went back further and were much wider than the stocking of his garden for teaching purposes. In 1761, on his first appointment, the flora of Scotland was still extremely poorly known: a scattering of localised records in the seventeenth-century publications of John Ray, and an unlocalised list of about 500 wild species in Sir Robert Sibbald's *Scotia illustrata* of 1684. Ray's records were repeated in William Hudson's *Flora Anglica* of 1762, a work Hope recommended to his students. As a botanical bibliophile Hope had several early local and regional British Floras in his library, including those of Ray and Thomas Martyn for Cambridge, G.C. Deering's 1738 work on the area around Nottingham, and Caleb Threlkeld's 1727 *Synopsis* of the plants growing near Dublin – the last of particular interest for its notes on medicinal and economic uses of plants and Gaelic names.

Hope was also well aware of what was happening in mainland Europe, and when showing his students the first part of the magnificently illustrated *Flora Danica* in 1763 pointed out (as reported by Timothy Bentley) that 'there was hardly a polished nation in Europe but had its Flora', and that he 'should be glad to see such a thing as a Flora Scotica'. Such works were not only academic – revealing floristic differences between geographical areas, and illustrating Linnaean taxonomy and nomenclature at local levels – but were part of the improvement agenda of documenting potentially exploitable natural resources on a national basis. But even at this stage Hope produced an excuse as to why he could not produce such a work by himself: 'it was too much for the Professorial Chair, as he that fill'd it must reside at Edinburgh'. There were, however, ways around this – first with 'assistance from some eminent men, [such as the Rev] Dr Walker of Moffat'; second with help from his own students. Thomas Arnold, reporting on this same lecture, given in April 1763, went so far as to state that 'the chief Design of [the lecture] ... was to make the Students lend him their assistance in collecting the Plants of Scotland for the forming of a Flora Scotica'. The hoped-for collaboration with Walker will be discussed later, as will Hope's cooling off, within three years, from the whole idea.

But in the meantime he offered a carrot in the form of a gold medal awarded annually to the student who collected the best *hortus siccus* (literally a dry garden – a herbarium). The medal was of great beauty, showing a cedar of Lebanon flanked by two lowly plants illustrating a quotation from the Book of Kings, 'A cedro ad hyssopum usque', representing the gamut of vascular plant diversity [fig. 22]. The designer of the medal is unknown, but each copy was struck in London at a cost of £1 8s 6d. As already noted, these medals would

Fig. 22. Gold medal, presented to J.E. Smith by Hope in 1781. Designed by an unknown London medallist, c. 1763. Linnean Society of London.

later be given as rewards to donors of valuable exotic seed, but the first three were awarded for herbaria. The first two went to American students, Arthur Lee of Virginia (1763) and Samuel Bard of New York (1764), followed in the third year by Adam Freer, a Scot. Freer was employed by Hope in the summers of 1765 and 1766 to collect plants, particularly on the west coast, and he intended to publish his own plant list, presumably a preliminary 'Flora Scotica', but this was prevented when Hope found him a good job as surgeon in the Factory at Aleppo in the Levant (now Syria; he later went to India). It was probably at this early stage of good intentions that Hope playfully wrote in a letter to Dr David Skene of Aberdeen, physician and naturalist: 'Nota if ever any of my young folks should publish a catalogue of the Scotch plants I shall insist upon it that the cryptogamia be published by you, so I give you full warning to prepare yourself'.

The specimens from Freer, Menzies [fig. 23], and more than forty other students, gardeners and gentleman-naturalists were added to Hope's growing herbarium, which in 1768 he catalogued in his own hand (with some later interpolations). This list survives and shows the extent and pioneering character of the exploration of the native plants of Scotland by the professor and his network ([Balfour] 1907; Fletcher & Brown

Fig. 23. *Delesseria sanguinea*, a red alga. Specimen collected at the Black Rocks, near Leith, by Archibald Menzies in 1777.

1970). The catalogue also notes which plants were drawn by James Robertson, but sadly neither these drawings nor the specimens themselves have survived. Along with this herbarium catalogue Balfour published a second document from Hope's collection, which has recently been identified by Professor Charles Withers as being in the hand of the Rev John Walker – this is a listing of plants around Edinburgh in the order of their flowering in the years 1764 and 1765: a 'Calendarium Flora of Edinburgh'. Such a phenological approach was then in vogue – A.M. Berger, a Linnaean pupil, had produced one for Uppsala that was republished, with a similar list for Norfolk (and an earlier one extracted from Theophrastus) by Benjamin Stillingfleet in 1761. It is somewhat surprising that Walker gave this list to Hope, as their relationship was not of the warmest – even after the gift of this floral calendar Hope could in 1769 complain to Richard Pulteney: 'Dr Walker has behaved with much more reserve to me than it is possible [William] Hudson could do towards you, and take my word for it my countryman is as close as yours is – I could give you such instances ...'. Walker became Professor of Natural History at Edinburgh in 1779, a post contested by Hope's disciple William Smellie.

Fig. 24. Page of the 1748 herbarium of Dr John Steedman, showing members of the family Umbelliferae. The lower names on the labels (in darker ink) appear to be in Hope's hand.

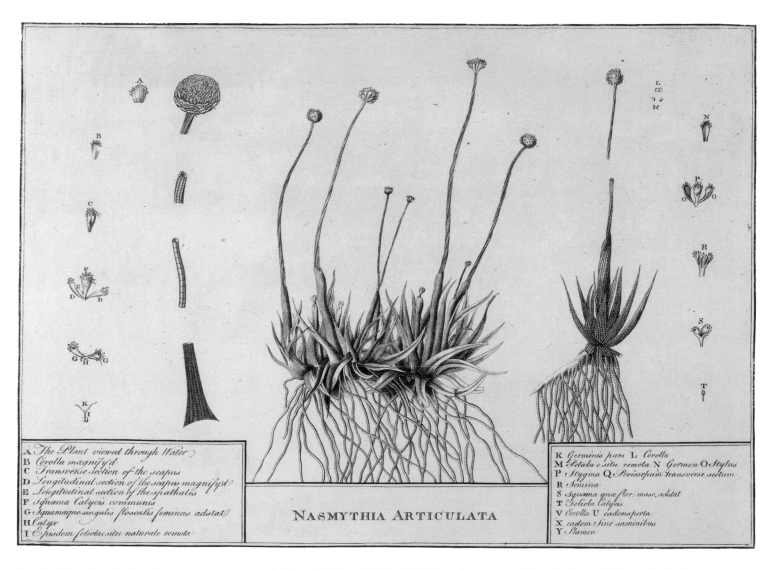

A The Plant viewed through Water
B Corolla magnify'd
C Transverse section of the scapus
D Longitudinal section of the scapus magnify'd
E Longitudinal section of the spathales
F Squama Calycis communis
G Squamaque singulis flosculis femineis adstat
H Calyx
I Episdem foliolaesitu naturale remota

NASMYTHIA ARTICULATA

K Germinis pars L Corolla
M Petala e situ remota N Germen O Stylus
P Stygma Q Pericarpium transverse sectum
R Semina
S Squama quæ flor. masc. adstat
T Foliola Calycis
V Corolla U eademaperta
X eadem sine saminibus
Y Stamen

Fig. 25. 'Nasmythia articulata'. Engraving by Andrew Bell, after a drawing by James Robertson, c. 1769. This is an early state of the engraving, bearing the provisional name that was never published by Hope. When he did publish the plate in the *Philosophical transactions* vol. 59 (1770) it bore the name *Eriocaulon decangulare*, but the plant is now known as *E. aquaticum*.

Withers (1991) and Eddy (2008) have both recently published on Walker (the latter concentrating on his geology), but neither has discussed his relationship with Hope – their interaction merits further work, as does Walker's botanical teaching.

With the loss of Hope's herbarium it seemed that there was no means of assessing the accuracy of his knowledge of British plants. But this has recently been somewhat remedied with the discovery that some of the annotations in a bound herbarium in the RBGE library compiled in 1748 by Dr John Steedman (or Stedman), a physician colleague of Hope, with specimens arranged according to Ray's system, are almost certainly in Hope's hand [fig. 24]. The plants included are interesting in their own right – both native Scottish species (including the Arthur's Seat rarities) and plants from the physic gardens of Edinburgh. Another interesting insight into Hope's interest in the Scottish flora has recently come to light in a study of the books from his library – in his own copy of James Sutherland's 1683 *Hortus medicus Edinburgensis*, Hope inscribed a 'List of Indigenous Trees', and an 'Addenda [of plant names] from Pennycuick's history of Tweeddale'.

Like Linnaeus, but in contrast to his contemporary Walker, Hope was not a great traveller – other than some visits to London his own major British travels appear to have been limited to the primarily horticultural tours of Scotland in 1765 and England in 1766. As already seen he explained this as due to the Professor's being tied to Edinburgh (which also happened to be the base for his lucrative medical practice). As explained in his 1763 lecture, and following the example of Linnaeus, Hope used carefully chosen apostles to pursue the utilitarian agenda of exploring and investigating Scotland's little-known natural resources, including plants. In 1767 he helped to

organise the journey of Andrew Thomson 'to research those parts of Scotland which he thinks have not been examined', instructing him 'to make diligent enquiry concerning the dyes which the natives employ ... and the remedies which they are in use of employing' (quoted by Emerson 1982). The Scottish excursions of Adam Freer have already been noted, but the individual who made the greatest contribution to this early, and rapidly expanding, knowledge of the Scottish flora was James Robertson, whom Hope himself referred to as his 'botanical missionary'.

James Robertson (d. 1796) was a gardener at Leith Walk from at least 1765 (on an annual salary of £11 14 shillings), and cousin of the head gardener John Williamson. In him Hope recognised 'a young man of promising genius though then illiterate' and took steps to educate and train him in botany and medicine. As with the other favoured gardeners, Menzies and Lindsay, Hope may not have entered Robertson on his own class lists, though there is someone of this name down for 1770 who attended other University medical lectures the same year – this may be him, but Robertson must surely have attended the Botany class more than once.

In 1767 Hope obtained a grant of £50 a year from the Commissioners for Annexed Estates to pay the expenses of sending Robertson as 'itinerant botanist' to collect plants in different regions of Scotland each year from April or May until the end of October. In this he was following in the footsteps of the Rev John Walker who had undertaken a similar expedition for the Commissioners in 1764. Robertson received payment for travel and living expenses and for the wages of a lad in each place he visited, which accounted for about £30 of the budget. In addition he got one shilling a day for himself while travelling, but this sum (£11 to £12) was evidently expected to keep him during the winter when he arranged his specimens and wrote his reports.

Robertson was probably the un-named emissary sent by Hope in 1766 to Hamilton, Ayr, Arran, and the 'chain of forts' to Inverness, but better documented are the expeditions made in the years 1767 to 1771, in the course of which he visited much of the Highlands, the Northern Isles, and several of the Inner and Outer Hebrides. His diaries or reports for four of these major excursions, together with a 'Tabular Flora and Fauna of the Islands' (Bute, Arran, Mull, Skye, Orkney, Shetland, and the Long Island – Lewis and Harris), were published by the late Douglas Henderson (a successor of Hope as Regius Keeper of RBGE) and Jim Dickson in 1994. These show an able, intelligent, and, by this time, highly literate young man who not only collected plants, but also made notes on birds, minerals and soils, and on the state of agriculture, gardening and fishing. Robertson's further explorations were curtailed as in January 1772 he was 'called away to the East Indies' as a Company surgeon, where, in China and India, he made enough money to buy an estate in Fife and build himself a mansion.

Two of James Robertson's most exciting discoveries, both from Scottish islands, were submerged aquatic flowering plants with almost microscopic flowers, though given his interest in seaweeds perhaps this is not surprising. The first, the spiral tasselweed, which he found in Orkney, Shetland and on the Long Island, he called 'Ruppia elongata'. Hope suggested to Linnaeus that it be called *Ruppia spiralis*, in reference to its long twisted peduncles, but this suggestion was not taken up and the same name was independently published sixty years later. Robertson's exquisite red chalk drawing of this plant is the only one of his drawings to have survived [fig. 26]. The second Robertsonian aquatic was the pipewort, now known as *Eriocaulon aquaticum*, which he found on the Isle of Skye in September 1768 (though it may have been found there four years earlier by John McPherson, later Sir John, Governor-General of India).

Fig. 26. *Ruppia spiralis*, spiral tasselweed by James Robertson, c. 1770. Red chalk.

The pipewort has proved to have an interesting distribution on both sides of the Atlantic, but Hope was more interested in its strange morphology – from the prominent transverse divisions of its leaves, and the rings around its worm-like roots, the plant body appeared to be made up of small tubes. In this respect he considered that it 'unites distant parts of nature' and sent a root to London to see if John Ellis thought it might be algal in nature. Specimens were also sent to Richard Pulteney, Bernard de Jussieu and Linnaeus for their views on the strange plant's identity and its taxonomic position. In 1769 Hope had a plate of it engraved by Andrew Bell from a now lost drawing by Robertson and sent copies to Pulteney (who was sworn to secrecy) and to Lord Bute. Bute was not discreet and passed it to his protégé John Hill, who under the name *Cespa aquatica* rather naughtily pre-empted Hope's own description of the plant. More cautious, Hope waited to hear the opinions of Jussieu and Linnaeus, and the name used in one of his only four scientific papers, a report of its discovery in the *Philosophical transactions* was *Eriocaulon decangulare* (Hope 1770). Hope was able to provide Linnaeus with a revised description of the genus *Eriocaluon* (Hope 1771), which he had at first taken to be new and intended to dedicate to his patron Sir James Naesmyth (some of the plates had been distributed with his preliminary name 'Nasmythia articulata')

[fig. 25]. This is probably the source of the myth that Naesmyth studied with Linnaeus, but when Hope asked Linnaeus that the plant, if new, be called 'Nasmythia' it is clear that there was no connection whatever. Hope had to explain to Linnaeus that Naesmyth was 'a most experienced and devoted student of botany here', which would not have been necessary if Linnaeus already knew him.

FLORA SCOTICA – A NATIONAL FLORA

In his first years as Professor, Hope undoubtedly intended writing a *Flora Scotica*. His ardour for this project, however, appears rapidly to have cooled, replaced by more pecuniary aims. In 1766 he told Richard Pulteney:

> I certainly shall never publish any catalogue of Scots plants unless it be in old age, for amusement, if it shall be my fate to live to old age – it were folly at my time of life to neglect real substantial [medical] practice to become the author of a catalogue.

Here is revealed Hope's love of money that will also surface in discussion of his students and their all-important fees. Rather than being piqued by the appearance of the Englishman, the Rev John Lightfoot's *Flora Scotica* in 1777 (as has sometimes been presented), Hope, on the contrary, was probably rather relieved.

Lightfoot was an Anglican cleric, holding a plurality of livings in Middlesex, Hampshire and Nottinghamshire, those in the last county the gift of the Dowager Duchess of Portland, whose chaplain he was, and whose collections of shells and plants he put into Linnaean order. Thomas Pennant, the Welsh zoologist and traveller, invited Lightfoot to accompany him on what was Pennant's second tour of Scotland in the summer of 1772. Despite contemporary, almost entirely unjustified, criticism *Flora Scotica* is a remarkable and (thanks to Pennant's patronage) handsomely produced work, and Alfred Slack (1986) performed a useful service in rebutting the criticisms and documenting the itinerary. Given the brevity of the four-month trip, the excellence of the resulting publication, which covers some 1250 species including a generous coverage of the cryptogams, was possible only through the extensive help Lightfoot received from others, and which he fully acknowledged – such as the Gaelic names and what would now be called ethnobotanical information provided by the Rev John Stuart of Luss, who accompanied the southerners on the western seaboard. Lightfoot's acknowledgement is worth quoting:

> I have pleasure first to mention with gratitude the name of Dr. *Hope*, the present celebrated professor of botany at *Edinburgh* who not only favoured me with the sight of his copious *Herbarium*, but permitted me the use of his notes and observations, the result of long enquiry.

Hope must also have shown Lightfoot James Robertson's drawings as the engraving of the edible seaweed, badderlocks (*Alaria esculenta*), in the book is based on one of them. Lightfoot acknowledged three other sources for Scottish records, the Dumfries-shire minister the Rev John Burgess, and two one-time students of Hope's whose herbaria he studied – John Parsons, who had taken Hope's classes in 1764 and 1765, and by the time of publication was Professor of Anatomy at Oxford; and Thomas Yalden, who studied medicine at Edinburgh after Lightfoot's visit, attending the Botany class in 1774. It is worth noting that Lightfoot's work remained the only national Flora until Hooker's *Flora Scotica* of 1821. But, over and above assisting Lightfoot, what Hope did was to establish a tradition of investigation of the native flora, using the most careful and comprehensive methods of observation and recording, which has continued to the present day.

Hope's interests in Scottish plants were far wider than the documentary (in words or pictures) and their potential in horticulture; and what can only be described as a proto-ecological approach is revealed in some of his lectures and unpublished notes. For example, long before Humboldt, he noted the phenomenon of altitudinal zonation, recognising three vegetational zones in Scotland – one characterised by:

the growth of grain such as wheat. The second one by the growth of heath, which, tho' it may intermix with the first, yet is perfectly distinct from the last where nothing but lichens grow, as on the tops of some of our hills.

Not surprisingly this was linked to an improvement agenda: 'it would be very useful if trees were brought in which could grow on these different regions, as by that [means] a vast quantity of waste ground would be made of use to the proprietor'. Doubtless he would have approved the extensive plantation of Sitka spruce and other North American conifers by the Forestry Commission in the twentieth century. Hope also noted the curious phenomenon of the occurrence of maritime plants, such as scurvy grass and thrift, on mountain tops. He was also interested in ideas of native versus introduced plants, and, following the Rev John Stuart, matters now considered under the heading of ethnobotany – including indigenous names. These more applied aspects of botany are also to be found in the work of Linnaeus and their economic motivation has been illuminatingly discussed by Lisbet Koerner (1999): from Hope's teaching they were taken up and extended by the best of his students in their travels – notably in India and the West Indies.

Fig. 27. *Astragalus uralensis* (now *Oxytropis halleri*), purple oxytropis. Engraving by Gabriel Smith, published in the Rev John Lightfoot's *Flora Scotica* (1777). First observed in 1767 by James Robertson between Harradale and Farr on the north coast of Scotland.

Hope and Linnaeus

In 1790 Hope's friend Richard Pulteney, a physician in Blandford, wrote that:

> the sexual system was received nearly about the same time in the universities of *Britain*; being publicly taught by Mr. Professor MARTYN, at *Cambridge*, and by Dr. HOPE, at *Edinburgh*. The adoption of it by these learned Professors, I consider, therefore, as the æra of the establishment of the *Linnæan* system in *Britain*.

Although this is not strictly true (Pulteney was unaware of William Cullen's Linnaean teaching at Glasgow in the late 1740s), this statement has often been repeated, and is probably one of the few facts generally known about Hope. Despite Britain's well-known clinging to the sexual system well into the nineteenth century, which history has deemed unfortunate, it is necessary to say something of Hope's links with the great Swede.

Hope and Linnaeus [fig. 28] never met, and their interaction was largely epistolary. The letters of Linnaeus to Hope have not survived, but the other side of the (Latin) correspondence is in the Linnean Society in London, and was translated by Morton and published in the first edition of this book. These letters largely concern the sending of botanical material to Linnaeus –

Fig. 28. Carl Linnaeus (1707–1778). Engraving by J.M. Bernigeroth of Leipzig, 1749, based on a portrait of 1748.
This engraving was used as the frontispiece of the first edition of *Philosophia botanica* (Stockholm, 1751).

including seeds (especially North American), specimens, and the prints commissioned by Hope of *Eriocaulon*, *Rheum* and *Agave*, but various taxonomic matters were also discussed. In 1765 he reported on how Bute's patronage had allowed him to combine and move the two old botanic gardens to a new site; he also told Linnaeus about some of

THIS PEDESTAL WAS ERECTED TO THE MEMORY OF SIR CHARLES LINNEUS BY PROFESSOR JOHN HOPE IN PRESENCE OF THE STUDENTS OF MEDICINE IN THE UNIVERSITY OF EDINBURGH. MDCCLXXVIII his star pupils. The following year Hope explained that he intended using Linnaean nomenclature in the next edition of the *Edinburgh pharmacopoeia*.

There were, however, more personal links, and three Linnaean pupils visited Edinburgh and met Hope. The first was the entomologist J.C. Fabricius, who came to Edinburgh to rescue his elder brother in 1767 – in his autobiography the younger Fabricius recounted that his brother 'introduced me to the heroes of medical science, and made me acquainted with Cullen, Gregory, Young, Hope in whose company we spent many evenings in scientific conversation'. Adam Kuhn (1741–1817) from Pennsylvania, but of German parentage, studied medicine and natural history at Uppsala from 1761 to 1764, and was probably Linnaeus's only American pupil. He came to Edinburgh for further medical study, and though he attended neither the Botany nor the Materia Medica classes, he asked Hope to send a copy of his 1767 MD thesis to Linnaeus via a ship to Gothenburg. Henric Gahn (1747–1816) was a Swedish student of Linnaeus who in 1767 defended a doctoral thesis on grasses, *Fundamenta agrostographiae*. Between 1770 and 1773 Gahn made a scientific tour of Germany, Holland, England and Scotland, and was sent back to Linnaeus bearing seeds from Hope with a note that 'Dr Gahn is much beloved by all here; he leaves behind a longing for his presence'. At least one student travelled in the other direction, and in 1776

Hope sent the young Dr Thomas Clarke (Clerk) to Linnaeus so 'that he may acquire from your own lips the latest developments in this Science'. He added that Clarke was 'proud of having met you and spoken with you on a former occasion', but where and when this earlier meeting had taken place is unknown, though it suggests that Clarke had previously visited Sweden.

Despite what he might have learned of Linnaeus through letters or pupils, Hope's greatest knowledge necessarily came through Linnaeus's publications. The books Hope owned are known from a list of his botanical library offered by his grandson's trustees to Isaac Bayley Balfour in 1899. Many of these are still at RBGE and bear Hope's bookplate with the family crest of a globe under a rainbow and the punning motto '*At spes non fracta*' ('yet hope not broken') [fig. 29], and from their marginalia it is apparent which were most heavily used. Hope owned eleven Linnaean titles, some in several editions, making a total of about 22 works. As is to be expected most of the great systematic works were there – from the *Classes plantarum* of 1738 (with its comparison of earlier classification systems, and Linnaeus's own pioneering natural one), to the classic *Genera plantarum* (5th and 6th editions), *Species plantarum* (1st and 2nd editions), *Mantissa* (1767, 1771) and the 10th and 13th editions of *Systema naturae*. Hope also had the two Scandinavian Floras – of Lapland (1737) and Sweden (both editions 1745, 1755), and the catalogue of

the Uppsala garden (1748). Given Hope's interest in the bibliography of botany it is not surprising to find that he owned Linnaeus's *Bibliotheca botanica* (2nd edition, 1751), and, for an understanding of the principles and practice behind Linnaeus's monumental edifice, *Philosophia botanica* (in the Vienna reprint of 1755), was essential. The final work was the first seven volumes of theses written by Linnaeus for his students to defend, collected under the title *Amoenitates academicae* (literally 'academic delights'), which, with their wide coverage of natural historical and floristic topics, were perhaps of greatest interest. Conspicuously Hope did not own the lavishly illustrated folio *Hortus Cliffortianus*, possibly beyond his pocket, for which no patron can have stepped in.

Fig. 29. Hope's bookplate. Unknown engraver, c. 1760. This impression is in Hope's own copy of the 2nd edition of Stephen Hales's *Vegetable staticks* (1731).

Fig. 30. Richard Pulteney (1730–1801). Engraving by James Basire, after a portrait by Thomas Beach, published 1804.
Royal College of Physicians of Edinburgh.

In the eighteenth century books were expensive commodities, and only the prosperous could afford to buy their own copies. Hope seems to have been generous in such matters and allowed favoured pupils such as William Smellie to use his teacher's library. He also ordered copies of standard works for Leith Walk (in 1767, for £1 9 shillings, copies of *Genera plantarum* and *Species plantarum*; in 1768, *Systema naturae* for 11 shillings); Hope was also almost certainly responsible for recommending the purchase of various books by Linnaeus for the University Library. But he went

further and issued (anonymously) two publications abstracted and simplified from Linnaean works, in order to make them available to his students at affordable prices. The first, *Termini botanici*, was a glossary of Linnaean terminology based largely on two works: the thesis of the same name defended by J.G. Smolandus; and the *Philosophia botanica*. The Edinburgh *Termini* was published in two editions in 1770 and 1778, but as these were in Latin, they were probably less useful to students than the (fuller) English version that John Berkenhout had thoughtfully published in 1764 and dedicated to Hope. The second 'abstract' from Linnaeus is Hope's (again anonymous) *Genera plantarum*, which also appeared in two editions. This contains brief generic characters arranged by class and order – those in the first edition (1771) are taken from the 12th edition of *Systema naturae*, those in the 1780 edition from J.A. Murray's '13th edition', his *Systema vegetabilium* of 1774.

Some of the urgent problems of eighteenth-century botany were linked with the discoveries coming to light from the rapid rise of Western colonialism. If the exploration of the immense variety of the world's flora was to yield scientific results, adequate technical methods for its description, naming, and classification were required. The practical genius of Linnaeus provided the necessary techniques – by establishing a precise

yet flexible terminology of plant description, and a stable and logical nomenclature. To this was added, in the sexual system, a method of classification, which, though explicitly artificial, was based on easily observed floral characters of high diagnostic value that were easy to use even by people with limited botanical knowledge. It is a measure of Hope's scientific understanding and its utilitarian application that he gave his students a thorough training in Linnaean methods long before their importance for botanical progress was generally appreciated. His teaching was responsible for the early adoption by almost all Scottish botanists of Linnaean nomenclature and descriptive terminology.

Despite his great respect for Linnaeus, Hope retained a critical and balanced attitude towards his work. For example, in the field of anatomy he firmly rejected, on the grounds of observation and experiment, the revival by Linnaeus of the old and persistent idea that the pith (*medulla*) is the most necessary part of vegetables. He also virtually accused Linnaeus of plagiarism, regretting that he had copied from Joachim Jungius 'without acknowledging it'. And while using it, Hope, like many others, complained of the new terminology of Linnaeus as 'unscientific & Barbarous' and of his not (in most cases) basing new plant names on their properties, which made it 'burdensome to the memory & almost impossible to retain them'. These, however, were relatively trivial matters and Hope's most explicit and revealing expression relating to his opinion of Linnaeus, long after the latter's death, was written in a letter to J.E. Smith in 1784 – he wrote: 'I am impatient to see the eighth volume of *Amoenitates*, although I did not admire Linnaeus so much as a philosopher as a naturalist and systematist'. This enigmatic statement should be interpreted with reference to Linnaeus's discussion in *Philosophia botanica* where he wrote that the science of botany was constructed by philosophers 'by clear deduction from rational principles'. As Stafleu (1971: 39) explained this should be read as 'by deduction from *a priori* principles', that is, an Aristotelian approach, and it was probably this to which Hope took exception. Though much further work is required to understand Hope's philosophical position, his approach was almost certainly that of induction from empirical facts rather than deduction from *a priori* principles or theories. The linking of his criticism with a desire to see the posthumous eighth volume of the *Amoenitates* (published the following year by Schreber in Erlangen) is probably not coincidental, since, as noted above, the *Amoenitates* contain much of interest to the natural historian, regardless of the system of thought underlying Linnaean classification. Hope's friend Pulteney

[fig. 30], in his survey of the publications of Linnaeus, aptly wrote that no summary of this composite work could give an adequate idea of the 'merit and excellent arrangement of the subjects in these volumes, which cannot but render them an agreeable and useful miscellany, and ornament to the library of every naturalist, philosopher, and physician'. The *Amoenitates* are still generally overlooked, but, to give a taste of some of the subjects covered of special interest to Hope, and used in his lectures, are dissertations on: the sexes of plants; the tastes and smells of drugs; hybrid plants; buds of trees; species monographs including rhubarb and *Betula nana*; the habitats of plants; instructions to travellers; dye plants; inebriating drugs; the drinks tea, coffee and chocolate ... and many more.

Hope had taxonomic discussions with Linnaeus over three plants. The two British ones (*Eriocaulon* and *Ruppia*) have already been discussed; the third was North American and involved Alexander Garden, an Edinburgh-trained doctor of Charleston, South Carolina, and correspondent of both Hope and Linnaeus. Early English settlers encountered a medicinal plant that they named the

Fig. 31. *Hopea tinctoria*. Specimen collected by Alexander Garden, presumably in South Carolina, 1767. This was the first genus named for Hope, by Linnaeus. It was soon realised not to be generically distinct and is now known as *Symplocos tinctoria*.

Fig. 32. *Hopea tinctoria*. Hand coloured stipple engraving by Gabriel, after a drawing by Henri-Joseph Redouté, from F.A. Michaux's *North American Sylva* (Philadelphia, 1865).

'Indian pink', which Linnaeus placed in the honeysuckle genus *Lonicera*. This plant was studied by Garden, who sent specimens and a report on its structure and medicinal properties to Hope. Both Garden and Hope realised that the plant could not be a *Lonicera* and that it might belong to the genus *Spigelia*, but were doubtful because it differed somewhat from *Spigelia anthelmia*, then the only species in the genus. Hope sent some of Garden's material to Linnaeus, who agreed, and in 1767 he transferred the species to *Spigelia*, where it has remained.

Alexander Garden was also the first botanist to propose that a genus be named after Hope, and made the suggestion to Linnaeus for a North American shrub, which the latter duly published in 1767 – with the single species *Hopea tinctoria* [figs 31, 32]. The French botanist L'Héritier subsequently concluded that the plant was not generically distinguishable from *Symplocos* so the American plant became *Symplocos tinctoria*. As the original *Hopea* had been 'lost', William Roxburgh in India felt justified in naming another genus, in a quite different family, after his old teacher. Under current botanical rules such reuse of a name is not permitted, but fortunately Roxburgh's genus has been allowed to stand (it has been 'conserved'), as, in addition to his *Hopea odorata* [frontispiece], the later genus now contains a further 103 species, distributed through tropical South and Southeast Asia, many of them valuable timber trees.

THE LINNAEUS MONUMENT

The most tangible reminder of Hope's regard for Linnaeus is the monument erected to his memory, a year after his death, in the Leith Walk garden. Writing to his friend Dr Maxwell Garthshore, Hope explained that:

> The design of the Monument I erected to Linnaeus was given me by Mr [Robert] Adams [sic] and it was executed by your acquaintance Mr Craig who gave the plan of the new Town.

This elegant and dignified memorial, in the form of an urn on a plinth bearing the words 'Linnæo posuit I Hope, 1779' can be seen in the present RBGE, to the north of the main range of glasshouses [fig. 33].

In Sir John Soane's Museum in London are three ink outline designs for the monument from the office of Robert Adam. One is more or less as Craig built it in Craigleith sandstone, a substantial cuboidal plinth with projecting, chamfered corners, topped by an ovoid, handle-less urn, with ornament reduced to a mask and foliated swags above the inset marble inscription – Hope originally had a worked-up version of this drawing in ink wash that he sent to Richard Pulteney, keeping only a rough copy. (Perhaps Pulteney contemplated using it as an illustration in his *General view of the writings of Linnaeus*.) Curiously this version is closely related to a design that Craig himself drew for a proposed monument to Lord Provost Kincaid in 1777 reproduced in Cruft (1995); given the dates, the question of who copied whom, or what the common source was, is an intriguing one. The monument shown in the second Soane drawing is similar, but slightly more elaborate, with a small, rather busty, female head and torso (*term*) at each of the upper corners of the plinth, and with a pair of volute handles on the urn – the worked up ink wash drawing for this version is still among the Hope drawings [fig. 34]. The third Soane version is altogether more remarkable and must have been ditched at an early stage on grounds of expense, as there is no wash version among Hope's collection. The basic form is the same, but at each corner stands a (life-size?) mourning figure carrying a down-turned torch, and, above the architrave, the urn sits on the intersection of two fish-scaled roof forms terminating in four segmental pediments, with acroteria at each corner. Stephen Astley (pers. comm.) comments: 'Robert Adam frequently used the tactic of offering something big and wonderful to entrance and entice potential clients. If they said "yes", he got to build something fantastic, and as he was usually on a percentage of total project costs, he got a bigger fee.' Clearly this is what happened here and Hope, no surprise, went for the cheapest option.

The Linnaeus monument is telling in several respects; not only is it a testimony to Hope's veneration for a colossus of botanical science, but as garden ornament it should be seen in the context of his ideas on garden design discussed in Chapter 2. On entering the garden from Leith Walk, through an unpromising door to the left of the gardener's house, the visitor would have seen the 140-foot facade of the conservatories sweeping westwards, but to the right, and much closer to the viewer, the monument would have formed a prominent punctuation mark at the centre of the 'Linnaeus' section of the garden [fig. 11 (19)].

Fig. 33. Monument to Carl Linnaeus, erected by John Hope in 1779. Designed by Robert Adam, made in workshop of James Craig.

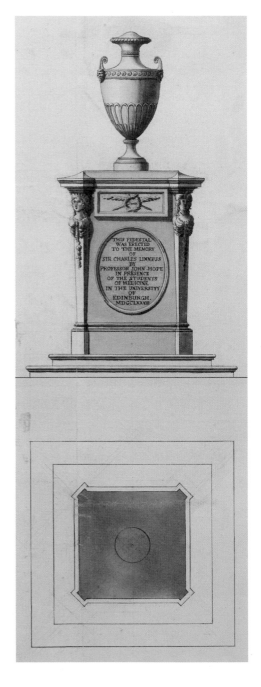

THIS PEDESTAL
WAS ERECTED
TO THE MEMORY
OF
SIR CHARLES LINNÆUS
BY
PROFESSOR JOHN HOPE
IN PRESENCE
OF THE STUDENTS
OF MEDICINE
IN THE UNIVERSITY
OF
EDINBURGH,
MDCCLXXVIII.

Fig. 34. Unexecuted design for the Linnaeus Monument by Robert Adam, c. 1778. Ink and ink wash over pencil.

THE LINNAEAN COLLECTION CROSSES THE SEA

Hope's teaching of the work of Linnaeus was to have a spectacular, if unforeseen, outcome through James Edward Smith [fig. 35], a wealthy and somewhat pampered youth, who attended the 1782 Botany lectures. Smith was already keenly interested in plants when, as a Unitarian, he chose to study medicine in Edinburgh; his letters home to his father in Norwich were enthusiastic about the help he received from Hope, who had:

> real goodness of heart, and is a man of the first consequence in this place: his behaviour was at first ... a little reserved; but botanical subjects opening the way, he became perfectly affable, and treats me with almost paternal tenderness.

At Edinburgh Smith, with a few friends, started an influential Natural History Society, of which Hope was elected an honorary member and the Rev John Walker, who let the society meet in the University Museum, an ordinary one. Smith set about preparing a herbarium with an eye to winning Hope's gold medal, towards which, in August 1782, he made a week-long excursion to Ben Lomond with the young Thomas Charles Hope and the

Rev John Stuart of Luss, so that alpine rarities unobtainable by most undergraduates could be included. The somewhat sycophantic strategy worked, and Smith's copy of the medal (one of only two known to have survived) is still in the Linnean Society [fig. 22]. Smith was at Edinburgh for two years then went to London for further medical studies with John Hunter.

It was at this point that the younger Linnaeus's executor J.G. Acrelius wrote to Smith's former Edinburgh room-mate Johan Henric Engelhart (by now also in London), offering the Linnaean collection of books, manuscripts and natural history specimens to Sir Joseph Banks for 1000 guineas. Smith happened to be breakfasting with Banks on the day he received the letter, 23 December 1783; Banks declined and passed the offer to his young guest, who persuaded his papa to pay up, and thereby became the owner of this priceless treasure, which arrived by ship from Sweden in October 1784, probably not, as tradition would have, pursued by a Swedish gun-boat [fig. 35]. In 1788 Smith founded his second natural history society, somewhat more ambitious than his first in Edinburgh – the Linnean Society of London. On Smith's death in 1828 he had instructed that the Linnaean collections be sold by his executor for the benefit of his family, so the Linnean Society had to raise £3150, which crippled them with a debt paid off only in 1861, but the collections remain with the Society to this day.

Hope was always keen to keep up to date with the latest botanical literature, and Smith acted as a useful go-between with London booksellers. In 1786 Smith translated from the Latin Linnaeus's *A dissertation on the sexes of plants*, which he dedicated to Hope. Smith's devotion to the sexual system long outran its dropping not only by Continental botanists, but by contemporaries including Francis Buchanan and slightly younger botanists, notably Robert Brown. But whatever the classification system he used, Smith's outstanding botanical monuments are the 36 volumes of *English botany* (1790–1814), with every species (flowering plants and cryptogams) illustrated by James Sowerby; his unillustrated *English flora* of 1824 to 1828 (which does include natural orders for the plants); his 3348 botanical articles for Abraham Rees's *Cyclopaedia*; and the completion, after John Sibthorp's death, of that most spectacular of all illustrated Floras, the *Flora Graeca*.

Fig. 35. James Edward Smith (1759–1828). Stipple engraving by William Ridley, after a portrait by John Russell. Published by R.J. Thornton in 1800.

Above the portrait is the plant *Smithia sensitiva*; the lower vignette shows the apocryphal 'Pursuit of the ship containing the Linnaean Collection by order of the King of Sweden'.

JAMES EDWARD SMITH M.D. F.R.S.
PRESIDENT OF THE LINNEAN SOCIETY

The Pursuit of the Ship containing the Linnean Collection by order of the King of Sweden

The botanical lectures of John Hope

One of Professor Hope's most fruitful and lasting contributions to the advancement of botany, and even more distinctively the expression of his unique personality and scientific gifts, was his botanical teaching, and the philosophical tradition it established and spread far beyond the national boundaries of Scotland.

Lecturing, however, did not come easily and in 1763 Thomas Arnold remarked on Hope's lack of facility in communicating his ideas verbally: 'for in his lectures he rather hesitates … but though his oratory has not gained admiration, his knowledge and learning have gained my esteem'. Of the same lecture Timothy Bentley reported even more graphically (if less grammatically): 'You would have sweat for him had you been there, he labours oft big with matter, which he does not want, but public speaking, seem to me to be a thing which is difficult'. This, however, was early in Hope's career and he clearly improved with experience, for J.E. Smith, who attended in 1782, could write 'I admire his botanical lectures; his delivery is agreeable, with as many "behoves" as Dr. Walker'.

The botanical lecture course was the most advanced of its time in Britain –

in the range of topics covered, its particular emphases (most notably physiology), the range of both historical and contemporary literature reviewed, the reporting of his own experimental work, his use of visual teaching aids and the training it provided for enquiring young minds. Hope may not have been eloquent, but he instructed his students in the fundamentals, and introduced them to the frontiers of contemporary research and speculation in a critical and open-minded way. His philosophical position, like that of his scientific contemporaries, shows the influence of Descartes, Locke and Hume, though this was expressed in simple, understandable language. For example, as an aside to a lecture on plant tissues he remarked:

it is the prerogative of man when he cannot be assisted by his senses to have recourse to his reason, & great are the discoveries that may be made this way, in this manner of investigation, our ground work must be a diligent observation of phænomena, & then let us exercise our reasoning faculty, taking care however to check a too exuberant fancy which might lead us astray.

In relation to the question of analogy between plants and animals, much discussed by naturalists at the time, Hope emphasised to his students (echoing the words of Locke) that decisive arguments must be drawn from experiments and not simply from analogy, and, moreover, that it was 'not sufficient to know the result of the Ancients' experiments; it is also necessary to perform the same things yourselves'.

Unfortunately Hope's own worked-up copies of his lecture notes have not survived, but transcriptions taken down by various students have. These, despite inherent problems, allow a fairly accurate picture of Hope's ideas to be drawn. It is as yet premature to make a critical assessment of Hope's teaching – such must await an edition of the lectures, on which the present author is currently working – but an outline of the course can be given here. There are five known sets of notes from the botanical lectures (a remarkably small number given the more than 1700 men who attended them), and a single set from the early ones on materia medica, details of which are given on page 101.

The course consisted of some 60 to 65 lectures held at the Leith Walk garden in May, June and July. Students were not pleased at having to stay behind for the early summer to attend them, but Hope did his best to get through them as quickly as possible – in one lecture assuring them that they would be free by harvest time

(surely earlier then than now?) and, in one year, that he wanted to finish his course before the start of the Leith Races! The course was in three parts: vegetation (mainly anatomy and physiology); classification (including terminology); and demonstration (of plants in the garden). For the lectures in the first section his method for a particular topic was to refer to the 'writers' (i.e., older authorities), to present experimental evidence, then to explain the 'uses' or 'purpose' of a given function (physiological or anatomical) in the 'œconomy of nature' – he would often discuss whether an animal analogy was helpful or not. James Cunningham's abstract of the lecture course is given in Appendix 3, but will be amplified here.

PART I. VEGETATION

After a preliminary definition of botany and a course outline, Hope started with four or five lectures on taste and smell. He strongly approved the use of the senses in acquiring knowledge: 'I believe Gentlemen that we derive more knowledge from the senses viz. the taste & smell, than from all books together'. Although animals had much stronger instincts he was 'persuaded we have innate power in a greater degree than is commonly believed'. He was fully aware of the problems of describing, quantifying and naming such subtle and complex properties, and critical

of older authors, especially Linnaeus, but nevertheless taught his students how best to prepare their organs of taste and smell in order to make the most accurate identifications possible. Though this section of the course was probably largely a hangover of Hope's materia medica days, taste and smell being largely of use in medical diagnosis, it must clearly have had some indirect value in honing the critical faculties of students in the analysis of plant affinities, and especially in the use of another sense, that of sight, to be discussed later.

Next came three lectures on similarities and differences between plants and animals, starting with the age-old problem of how to define a plant and their relation to the animal and mineral kingdoms. He quoted and discussed the Linnaean aphorism that minerals have growth, vegetables have growth and life, and animals have growth, life and feeling, adding his own teacher Alston's picturesque view that a plant was an inverted animal, and John Hunter's related view that the main difference between plant and animal was that an animal had an internal receptacle for food. There followed a discussion of the 'purpose' of plants in the economy of nature – as food for animals, in preparing soil, in redistributing water and electrical fluid, and in cooling the

Fig. 36. *Castanea sativa*, an ancient chestnut tree, the 'Castagno dei Cento Cavalli', on the slopes of Mount Etna, Sicily. By Andrew Fyfe, c. 1780, probably based on a drawing made in the field by Daniel Rutherford. Watercolour and ink.

This tree, which is several thousand years old and has been legally protected since 1745, still stands. It was known to Hope from Patrick Brydone's published travel diaries of Sicily and Malta.

atmosphere. Here Hope discussed the work of Joseph Priestley and Jan Ingen-housz on the 'restoration of vitiated air' by plants in light. A topic that clearly intrigued Hope was the variability in lifespan exhibited within the plant, as compared with the animal, kingdom, and a particular interest in ancient trees. He discussed the yew at Fortingall in Perthshire, as recorded by Thomas Pennant and Daines Barrington, but almost certainly seen by Hope during his 1765 tour. Another favourite was the 'Castagno dei Cento Cavalli' on the slopes of Mount Etna, which he didn't see for himself, but which two students (Daniel Rutherford and Thomas Bowdler) independently investigated and measured for him,

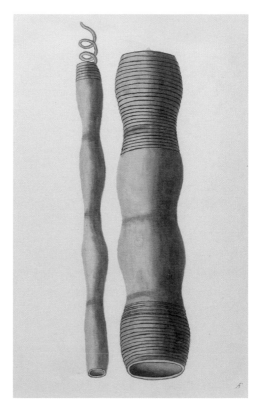

Fig. 37. 'Vasa aeria'. Adapted from a figure on Plate 5 of Marcello Malpighi's *Anatome plantarum* (London 1675). Ink wash drawing by Andrew Fyfe. This depicts two xylem elements of wood, the narrower one showing spiral thickening.

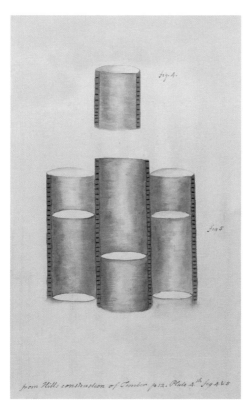

Fig. 38. Microscopic views of a tangential section through cells of the outer bark ('rind') of scarlet oak (*Quercus rubra*), copied from Plate 4 of Sir John Hill's *Construction of timber* (London 1770). Ink wash drawing, probably by Andrew Fyfe.

Anatomy

Inserted at this point in the 'Copyist' set of notes comes a fascinating account of how to collect, preserve and mount herbarium specimens, but the next major topic to be treated was plant anatomy, which occupied about six lectures. The authorities cited were Nehemiah Grew and Marcello Malpighi from the previous century, and from his own, Henri Louis Duhamel du Monceau, Charles Bonnet and N.A. Mustel in France, and in Britain the Rev Stephen Hales and Sir John Hill. Hope went into great detail about the structure of wood, in preparation for the next, physiological, part of the course. This cannot be summarised or analysed here; suffice to say that despite great advances then being made in microscopy, especially by Hill, it was impossible to obtain a clear picture of the fine structure of the vascular system of plants (and Hope was especially interested in the woody variety). He had no knowledge of the cambium, or the distinction between xylem and phloem, and the description of, and the terminology available for, the various elements of the vascular system ('vasa propria', 'vasa æria', various fibres, 'integumentum cellulare' ...) [figs 37, 38] and the fluids transported (sap, 'succus proprius') were inadequate, and the account accordingly somewhat confusing. Hope started by describing the scales that protected the 'infant'

and of which Andrew Fyfe made a drawing probably based on Rutherford's field sketch [fig. 36]. Determining whether or not its present appearance, a ring of seven trunks, had once formed a single tree was an opportunity to emphasise the need for care in observation and interpretation – as also was the manner of estimating the age of trees from

historical record or by extrapolation from the measurement of annual increments. This section ended with a discussion of the means by which plants 'engender' or reproduce themselves – sexual and vegetative – and the applied use of the latter in techniques such as cuttings, budding and layering, the topic of sex being returned to in greater detail later on.

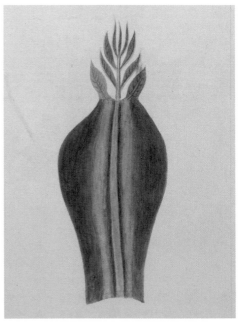

plant bud (the 'gem') [fig. 39], then proceeded to a detailed account of the bark, wood and pith. Contra Linnaeus, he realised that pith or medulla was not the site of growth, but instead emphasised a structure called by Hill the 'corona' or 'circle of propagation'. In *Bocconia* Hill correctly placed this as the outer ring of vascular bundles, but in trees such as oak he wrongly ascribed this function to a zone between the pith and the wood.

Physiology

There follows what is really the core of Hope's teaching, the area that appears to have fascinated him most deeply: approximately ten lectures on the 'private œconomy of vegetables' – by which he meant growth and physiology. In this his great mentors were the French Academicians, especially Duhamel du Monceau, and, above all, Stephen Hales. The copies of Hales's and Duhamel's works are by far the most heavily annotated books in his library, and, with the help of his gardeners, Hope went to great trouble to repeat many of their experiments in the Leith Walk garden. At least by demonstration

Fig. 39. Bud scales of the ash (*Fraxinus excelsior*). Anonymous watercolour.
This series shows the transition from simple, undivided scales into leaf-like ones.

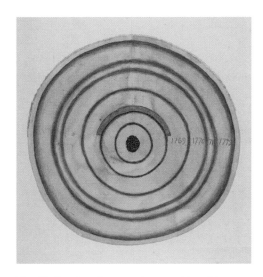

Fig. 40. Drawing of an experiment on 'the origin of the new stratum of wood', showing a lead plate inserted beneath the bark of a willow tree in 1768, and cut down 1773. Watercolour.

students were thus introduced to Hales's experimental and quantitative methods for measuring transpiration and water movement in plants and for observing growth. The first lecture was on the manner of growth in length and girth, illustrated with drawings of Hope's own experiments. He stuck evenly spaced pins into shoot and root tips, which showed that growth in length usually occurred only at the apex, but that there were differences between species, and between roots and shoots [fig. 41]. Growth in girth occurred by the addition of annual strata, and the question was whether these came from the bark or the wood.

Hope did not know of the existence or role of the cambium, and the experiments of Duhamel that he repeated, by inserting lead plates between the bark and the wood [fig. 40], were inconclusive, though they tended to show that the increments were added from the inside of the bark. Hope realised that this property accounted for the success of budding, and also allowed for the fusing of tree branches, which he spectacularly exhibited in the fusion of three willow trees (an experiment copied from Hales), which also demonstrated the lateral movement of sap [fig. 21].

What Hope termed 'general functions' were next considered – 'perspiration' (transpiration) and absorption – two lectures devoted to each, broken down into quality, quantity, causes and circumstances affecting these functions, such as heat and light, the organs that performed them, ending with their 'uses' in the system of nature. Hope distinguished between 'sensible' (i.e., having a distinct smell or colour) and 'insensible' perspiration. The lectures on absorption again quoted extensively from the experiments

Fig. 41. Drawing of an experiment demonstrating apical growth of a shoot of the 'plane' (*Acer pseudoplatanus*), made in the Leith Walk garden 1772.
Ink and watercolour.

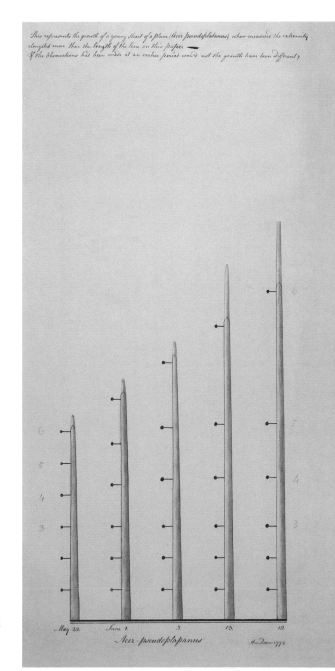

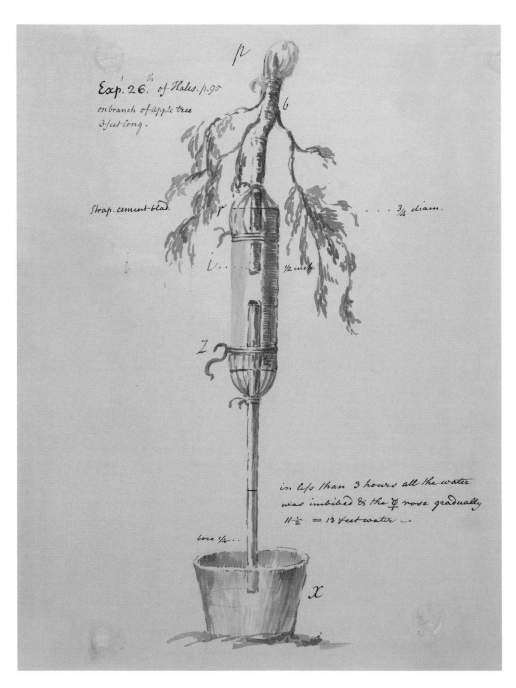

of Hales [fig. 42], and a famous experiment of Van Helmont, showing that a willow increased in weight if fed on water alone. Three lectures on sap followed, enquiring into its qualities, motions, and course. The burning question in this field was whether plants exhibit a 'circulation' analogous to that of the blood in animals. Hope gave a balanced account of the evidence and summed up in favour of Hales, who rejected circulation in plants. Hope himself made numerous experiments to reproduce and convince himself of Hales's results, for example by applying ligatures, and by making a spiral incision in [fig. 43] or ringing [fig. 44] bark, to see where, and under what conditions, 'bleeding' of sap took place; he also inverted tree branches to see if its motion was uni- or bi-directional; and conducted experiments such as the tree grafting one to see whether or not oblique (i.e., lateral) motion took place. Following French investigators Hope also made observations on sap flow using coloured liquids, and demonstrated their entry through the roots and into

Fig. 42. Copy of Plate 6 from Hales's *Vegetable staticks* (1727). Ink wash, possibly by Andrew Fyfe. This shows Hales's Experiment 26, where the power of perspiration of an inverted leafy branch of a golden pippin apple inserted into an 'aqueo-mercurial gauge' has the power to transpire the water from tube i, raising the mercury in tube z. Hope had this drawn to 'show the suction of ... branches'.

the wood. He gave a critical and cautious account of the use of coloured fluids in such investigations, concluding that experimental difficulties had not yet been overcome.

After this extensive discussion of absorption and perspiration came a lecture on the question 'do vegetables select their juices?', and on the green colouration of plants. The question arose since clearly some internal liquids of a plant differ from water in the soil – was this was due to selection of what was absorbed, or a transformation within the plant by means of 'internal energy'? Related to this was the cause of the unique green colouration of plants – was it, as an author in the *Philosophical transactions* had recently suggested, caused by an uptake of iron? Hope was aware of the need of light for the development of the green colouration and that in darkness plants lost their colour and sensible qualities, and he did experiments to see if this property of daylight could be simulated by candlelight. The nutrition of green plants by the assimilation of carbon from atmospheric carbon dioxide in the light (photosynthesis) was not fully revealed until after Hope's death, but he pointed out that plants absorb light and do not part with it, which might suggest a nutritive function for light as suggested earlier by Hales.

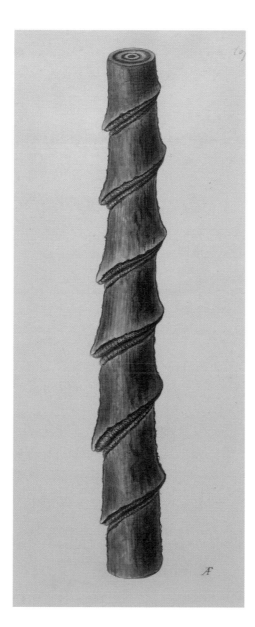

Fig. 43. Drawing of an experiment to show the effect on the motion of sap by making a spiral incision in the bark of an ash tree. Ink wash drawing by Andrew Fyfe, c. 1775.

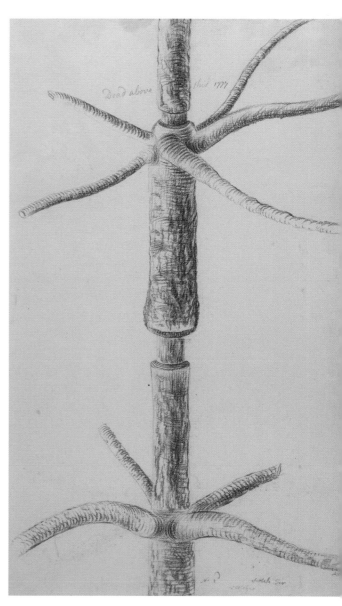

Fig. 44. Drawing of an experiment to show the effect on the motion of sap by making 'circular decortications' (ringing) in the bark of Scots pine. Red chalk drawing by Andrew Fyfe, c. 1777.

The motions of vegetables

The discussion of the effect of stimuli such as light led to a consideration of the 'motions of vegetables', to which Hope devoted probably two lectures. In the first he described what are now called geotropism and phototropism, and the different responses of roots and shoots to light and gravity. The latter he demonstrated by means of (unintentionally picturesque) experiments suspending pots of mint, *Tagetes* and *Asclepias* upside-down, to see how the plants would reorientate themselves, if left alone [fig. 45] or with a weight tied to them. What has been claimed as Hope's most original experiment, in which a pot of woodruff (*Asperula odorata*) was suspended, and a mirror used to determine the interaction of light and gravity on growth, proves not to have been of Hope's own devising [fig. 46]. Three of the four drawings of the experiment are signed by John Lindsay, and fortunately Francis Buchanan recorded Hope's own statement that these 'Experiments [were] contrived and executed by John Lindsay, an ingenious man whom I have before mentioned'. Neither Hope nor Lindsay published this experiment, which pre-dates by nearly a century any similar work on

Fig. 45. Drawing of an experiment showing the negative geotropism of an inverted shoot of *Tagetes* under illuminated conditions. By John Bell or Andrew Fyfe, c. 1780. Watercolour and ink.

Fig. 46. John Lindsay's experiment on competing effects of light and gravity. Drawings by John Lindsay, c. 1780. Ink, charcoal and watercolour.
These show the power of light (via a mirror) to overcome negative geotropism of a shoot of *Asperula odorata* under dark conditions – as Francis Darwin expressed it in 1909, the 'victory of heliotropism'. (The other two drawings in the series are controls, showing negative geotropism of the shoot under light and dark conditions.)

Fig. 47. Sleep movements of leaves. Possibly by Andrew Fyfe, c. 1780. Ink wash.
This shows the diurnal position (left), and the nocturnal position (right), with upper surfaces of leaf together 'face to face'. Although labelled *Tamarindus indicus*, it is probably a species of *Senna*.

the interaction of plant tropisms. Also discussed were sleep movements of the leaves of *Oxalis* and various legumes (clover, 'tamarind', etc.) [fig. 47] as described by Linnaeus – and their analogies (or not) with animals, a question also discussed in connection with 'irritability'. Under irritability were included movements not due to a purely mechanical cause (such as the bursting of anthers or fruit capsules), but in response to a specific stimulus, such as those elicited by touching the staminal filaments of *Berberis*. Here also came a discussion of the curious leaf movements of the *burrum chundalli*, which were not in response to touch, and those of *Mimosa pudica* and Venus's fly-trap (*Dionaea muscipula*), which were.

Sex and hybridisation

In 1694 Rudolph Jacob Camerarius had published the first experimental evidence that in flowering plants the pollen functions as the male fertilising element. This proof of the existence of sex in plants caused a revolution in botany by denying the long accepted view of Empedocles, reiterated by Aristotle, that plants lack sex and sexual parts. Although some experimenters confirmed the conclusions of Camerarius, others obtained negative or confusing results, so that for a considerable time it was possible to have legitimate doubts. There were also some, including J.P. de Tournefort, who agreed that stamens and pollen were needed for

seed formation, yet held that their function was not sexual but to excrete or divert material hindering the proper course of seed development. Nevertheless, by 1750 a majority of botanists, including Boerhaave of the older generation, and the Jussieu brothers and Linnaeus among the younger, accepted the new doctrine [fig. 48].

In introducing this still controversial topic to his students (in about four lectures) Hope did not conceal his own opinion that the sexual function of the 'farina' (pollen) was well established by 'a great variety of experiments'. He gave an admirably critical account of the evidence, referring to two sources that argued

Fig. 48. *Vallisneria spiralis*: female (left) and male (right) plants. Copied from Plate 10 of P.A. Micheli's *Nova plantarum genera* (Florence, 1729). Ink wash drawing by Andrew Fyfe, c. 1780.

Micheli considered these plants to belong to different genera, rather than being male and female of a single species. Erasmus Darwin had this same plate copied for his *Botanic garden* (1789).

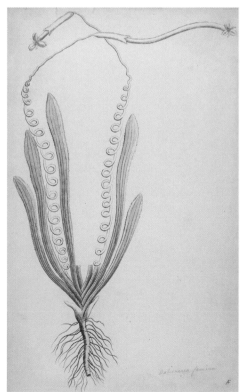
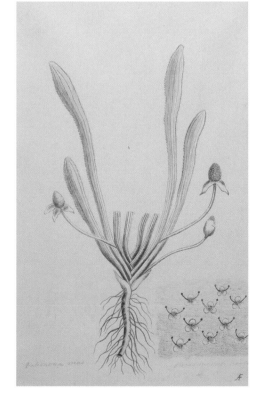

against plant sex – the article by his own pupil William Smellie in the *Encyclopaedia Britannica*, and the experiments of his teacher Charles Alston, but pointed out how the latter's experiments could have been misleading. In response to Alston's finding that female plants of normally dioecious spinach set a few fertile seeds in the absence of male plants, Hope quoted observations of his head gardener John Williamson that occasional staminate flowers appeared on pistillate plants – 'so you see the difficulties of experimental proof' [fig. 49].

The importance of pollen in various ways beyond its role in 'generation' was noted by Hope, and he even hinted at its potential diagnostic value – 'each species of plant has its pollen of an uniform shape ... each species has a different figure from the other species'. Hope also knew about insect pollination, noting that 'insects feed on one species of plant only, they therefore communicate the farina'. To his students he cited recent observations published in the *Philosophical transactions* made by Padre de Torri of Naples, with the

help of high-powered microscopes that were examined by the Royal Society's instrument maker Henry Baker. But not content with second-hand information from literature, Hope, in 1774, got two students, Thomas Clerk (Clarke) and James Kerr, to make drawings of their own microscopic observations [fig. 50].

Pollen blows around in the wind and does not always land on stigmas of its own species, which in flowering plants not infrequently leads to the production of hybrid plants. Hope knew of Linnaeus's work on hybridisation in *Tragopogon*, and other early work in *Zea*, but he also discovered hybrids of his own. Intriguingly the first were in the rhubarb genus *Rheum*, in which he had a particular interest:

> A curious thing has happened in this garden – a [new] plant of the *Rheum* kind made its appearance. Its leaves were not so much cut as the *Rheum palmatum*, but not entirely whole however, approaching nearer to the English [probably *R. rhaponticum*], but then it has the tall stem of *R. palmatum*. This bastard kind was in considerable quantity in a plantation both of mine & of the Duke of Atholl's.

Fig. 49. *Spinacia oleracea*, spinach, showing male (left) and female (right) flowers; and the hermaphrodite (centre) ones observed by John Williamson. Ink wash. Note. Two anthers are missing from the male flower.

Gardeners had, of course, been long interested in hybridisation from a practical point of view and John Williamson experimented with crossing the oriental with the opium poppy; Hope reported this work to the younger Linnaeus (without mentioning who had done it).

While the various student notes for the first part of the course are relatively consistent and comparable, they diverge more widely for the next two. Because of their practical nature they were doubtless harder to take notes from, but in these parts Hope probably also varied the order and content from year to year, not least for practical reasons of season, and what plant material was available. In Cunningham's year, before explicitly commencing the 'Classification' part of the course, Hope seems to have given four introductory lectures on floral structure, starting off with a classically simple spring flower, the tulip and gradually working through more complex ones to the legumes. The students dissected their own flowers, using a 'needle with three sharp corners', a flattened double-edged blade ('ancipites') and a magnifying glass.

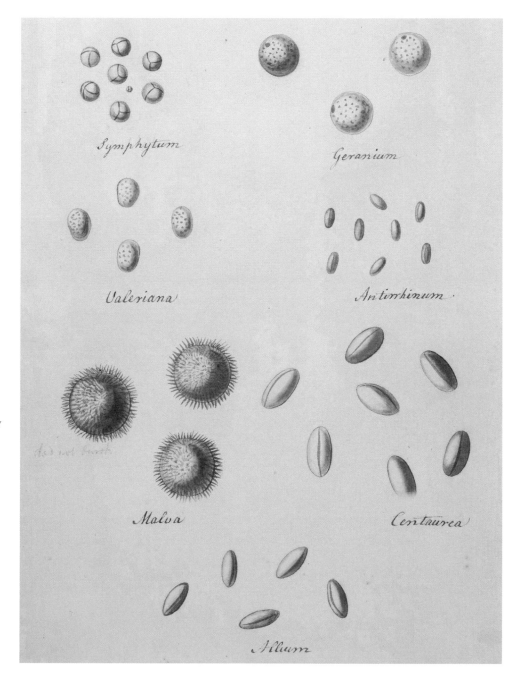

Fig. 50. Drawings of pollen from species of seven different genera, copied as a teaching drawing from originals made with the aid of a microscope by James Kerr in 1774.

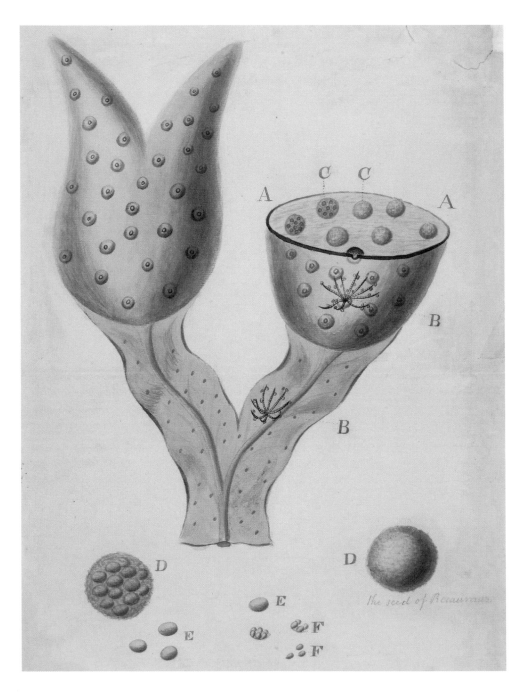

PART II. CLASSIFICATION & PART III. DEMONSTRATION

The overlap between the second and third parts of the course ('Classification' and 'Demonstration') is even more blurred than that between the first and the second. The two together appear to have represented about half the total course, around thirty lectures, but it is hard to say exactly how many lectures Hope considered were devoted to 'Classification', though probably around twenty, leaving between half a dozen and a dozen to 'Demonstration'. For the same reason that the students found these sessions hard to write up – too many facts and very little narrative – only the barest outline of these systematic parts of the course can be given here.

Hope made the important distinction between nomenclature and classification: the purpose of his teaching was not just to memorise names, but, by presenting a 'history of the several Plants in a scientific view', to understand them

Fig. 51. *Fucus vesiculosus*. Fertile branch and sexual organs. Ink wash drawing, possibly by Andrew Fyfe, c. 1775–80.

The left-hand branch shows a whole receptacle (probably female); the right-hand a receptacle in cross section revealing spherical conceptacles in which the sex cells are produced. Based partly on a plate published by R.A. Ferchault de Réaumur in 1711, but the microscopic details may have been drawn from life.

within the context of a classificatory system. Despite his reservations about the Linnaean sexual system, and the 'barbarity' of its terminology, he had the sound practical sense to use it in his course as the structure through which to examine plant diversity, and to teach its descriptive terminology. But before getting down to 'the scientific division of vegetables' (grinding through the Linnaean classes) Hope introduced Linnaeus's large-scale 'natural division of them into great families, which are 7 in number': Fungi, Algae, Musci, Filices, Grasses, Palms, and [Other] Plants. He regretted the last – what would now be called a 'dustbin' group – and made 'an observation unfavourable to the progress of Science when I say that these are many, & contain most of the Plants useful in Diet or in Medicine'. But two points can be made about this classification. The first is that it shows Hope's preference for groups with a uniquely defining set of characters, even when this meant unsatisfactorily leaving 'the rest'. It also gives prominence to the cryptogams, in which Hope took a particular interest. So, though he would not have realised it, he started his classification lectures with something that would now be understood as an evolutionary approach, going briefly through the non-flowering plants.

Under fungi he not only showed students how to make spore prints, but also devoted some time to fungal diseases of higher plants such as mildew of wheat and ergot of rye, citing Linnaeus and Boissier de Sauvages. Doubtless at least in part because of Edinburgh's proximity to the sea, Hope had a more than passing interest in algae, which he discussed in letters with John Ellis, then the great British authority on the group. Three pupils were asked to collect specimens, Archibald Menzies [fig. 23], Andrew Duncan in Fife, and James Robertson on the west coast. Hope was familiar with the pioneering algal work of the French natural philosopher René Antoine Ferchault de Réaumur, and showed *Ulva*, *Fucus* [fig. 51] and other marine algae to his class. Mosses interested Hope not least for their power of withstanding desiccation (in which they were compared with snails), but he had large drawings made of *Polytrichum*, *Bryum* and *Splachnum* [fig. 52], and ordered Hedwig's latest works from London booksellers. With the aid of magnifying glasses Hope had his students watch the discharge of globulets (spores) from globules (sporangia) in ferns, and he discussed the as yet unproved supposition that the globules corresponded to seeds – work greatly developed by his student John Lindsay in Jamaica.

Having briefly covered plants with hidden fructification, the cryptogams (which formed Linnaeus's 24th class), Hope pointed out that the other plants had stamens and pistils, usually in the

Fig. 52. *Splachnum rubrum*. Anonymous watercolour, c. 1780. This drawing is adapted from two small engravings on Plate 83 of J.J. Dillenius's *Historia muscorum* (1768), where it was called 'Muscus Norwegicus umbracula ruberrimo insignitus'. This is an Arctic species that does not occur in Britain. As Hope noted, the calyptra of the capsule is missing.

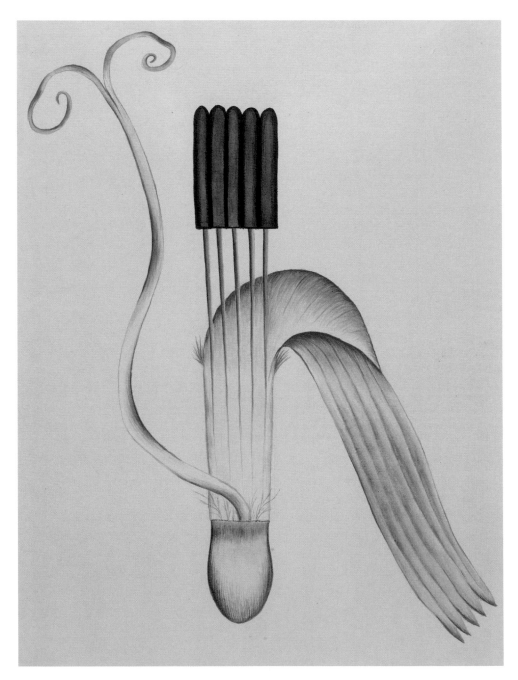

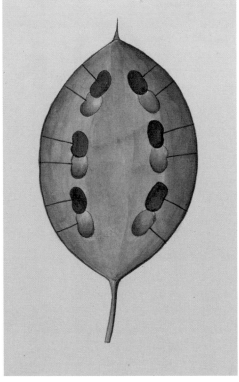

Fig. 53. Hermaphrodite, ligulate floret ('*flosculus*'), with anther tube split to free the style, of a member of class Syngenesia, order Polygamia Æqualis such as *Leontodon*, from a series of 'Syngenesious Flowers'. Anonymous watercolour, c. 1780.

Fig. 54. *Lunaria annua*, fruit. Drawn from nature by Agnes Williamson, 16 June 1783. Ink wash.

'The *Lunaria* is one of the class Tetra-Dynamia. Its pericarpium has two valves and the Seeds are attached alternately to both sides of the Suture, and is of the species of Pericarpium called Siliqua'.

Fig. 55. *Magnolia grandiflora*, fruit. Copied from Catesby's *Natural History of Carolina ...* (1738 or 1771) by John Bell, 19 June 1780. Watercolour.

'Every seed is connected to the Placenta ... *Magnolia grandiflora* is remarkable for this – it is hung by umbilical ropes from the outside'.

Magnolia grandiflora June 19th 1780

same flower (hermaphrodite) [fig. 53], but placed in different classes if male and female were separated on the same (Monoica), or different (Dioica), individuals. The remaining classes were based either on the way the stamens were joined together or on their absolute number. Hope then turned his attention to the female parts of the flower – the terminology of the various sorts of fruit ('pericarp') [figs 54, 55], and the number of styles in the flower as used by Linnaeus for defining his orders. Gradually more and more terminology was introduced, moving from sexual to vegetative parts, all demonstrated with living examples or sometimes drawings.

Natural classification

Like Linnaeus himself, Hope realised the limitations of the sexual system – in particular, Hope pointed out that the first eleven classes were based entirely on stamen number, which was notoriously variable (and therefore unreliable), and that under the sexual system many natural orders (such as grasses and lilies) were 'greatly broken through'. A major reason for preferring a natural system was for Hope a practical one, as medicinal properties were more likely to be shared by other members of a natural group. In Hope's words: 'nobody can acquire a proper idea of materia medica without attending to the natural orders', a proposition taken to heart by the young Andrew Duncan.

In fact one of the most fundamental questions in systematic botany during the eighteenth century, with important theoretical and practical implications, was the emerging concept of a natural classification of plants based on the 'sum of their affinities', as Robert Brown defined it early in the next century. In the early 1700s the natural classification of plants was already recognised by certain botanists as an immediate scientific problem, and the methods by which such a classification ought to be established were discussed. Linnaeus himself, in spite of the wide acclaim of his sexual system, said that the 'natural method' should be the ultimate goal of botany, and in his *Classes plantarum* of 1738 he had made an important step towards this by placing the genera in 64 groups or 'natural orders'. The principles to be employed in forming a natural classification were studied most deeply by Bernard de Jussieu at the Jardin du Roi in Paris, whom Linnaeus had visited in 1738, and who by 1747 was using multiple affinities to improve the Linnaean natural orders. Jussieu worked out his 'natural method' on the ground, in the physical arrangement of the beds he laid out in the royal garden of Louis XV at the Trianon Palace between 1759 and 1774, at which point the plan was adopted in the enlarged Jardin du Roi. Interestingly, in 1777 Hope was sent a manuscript catalogue of the Jardin du Roi, made by the landscape gardener Thomas Blaikie, with the species arranged under 14 Jussieuan classes. Bernard died in 1777 and it was left to his nephew Antoine-Laurent to publish the final scheme in his *Genera plantarum*, which did not appear until 1789, three years after Hope's death.

When the young John Hope studied under Jussieu in 1748/9 he must have been fired by the intellectual appeal of the new ideas on classification and the prospect of the research required to apply the 'natural method' to the whole plant kingdom and the flood of new species then being discovered. But it should not be forgotten that other botanists in Europe were working on natural systems as an improvement on earlier classifications, including those based on the form of the corolla of the flower by Rivinus and Tournefort. Even in France there were workers other than Jussieu, and Hope owned a summary of the natural classification by Louis Gérard from Provence. When Michel Adanson published his *Familles des plantes* in 1763 Hope must have ordered a copy immediately, for in October 1764 he had 'got Adanson some time ago'. In northern Europe there were also botanists working on non-Linnaean systems: in Germany those of C.G. Ludwig and Albrecht von Haller were based on Rivinus, but in Denmark G.C. Oeder's was influenced by Jussieu, and in Austria G.A. Scopoli worked on a natural system similar to that of Linnaeus, but he was later influenced by Adanson. Hope knew of all of this contemporary work. In addition to the relevant Linnaean works on the natural orders, including the scheme Linnaeus produced with Adrian van Royen for the Leyden garden catalogue, Hope had copies of the works of all these other authors in his library, and made notes comparing their systems with each other and with those of Sir John Hill and even Lord Bute's.

Given Hope's passion for natural classification, which he thought was 'to the botanist as the compass is to the mariner', it is somewhat surprising that it does not feature more extensively in the student lecture notes. There is one exception: those of Francis Buchanan, who gives a fair summary, and reproduces the keys displayed by Hope showing the system at the point it had by then reached. The original large-format versions of these keys have also survived [fig. 56], and with these, a recently discovered notebook (which assign genera to the orders), the notes in Hope's unpublished papers and details in a letter to Pulteney of 1777, it will, in due course, be possible to give a much fuller account of the 'little aid to lead to the natural method' on which Hope was working and hoping to publish 'soon' in 1781. At the highest level in Hope's hierarchy were seven classes:

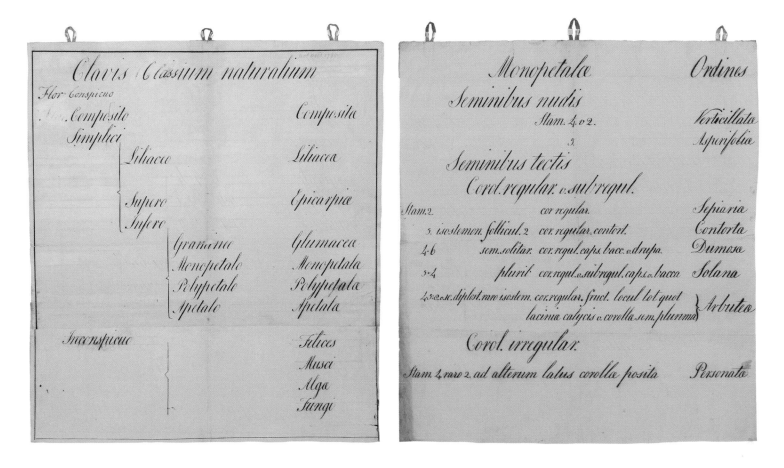

Fig. 56. Charts relating to Hope's natural system, shown to his class c. 1780. Left: Key to Hope's seven natural classes of phanerogams (and four of cryptogams). Right: Key to the eight natural orders of his class Monopetalae.

Note the loops showing how the charts were suspended for display during lectures.

Compositæ (with condensed, compound inflorescences); Lileæ (lilies); Epicarpii (with inferior ovaries); Graminæ (grasses); Monopetalæ (flowers with fused corollas) [fig. 56]; Polypetalæ (flowers with distinct petals); Apetala (flowers lacking petals). He then gave keys breaking down each of these Classes into 43 Natural Orders [fig. 56] – many with names used earlier by Linnaeus, though usually comprising somewhat different genera, but many more or less corresponding to Jussieuan and/or currently recognised plant families.

Demonstration

The lectures on 'Classification' merged into 'Demonstration', the last part of the course, the main purpose of which was to show living examples, in garden and greenhouse, of plants displaying features, properties, or functions, discussed in the first two parts. At one point the Professor said that he would not dwell on medicinal properties, which from 1768 were the territory of Francis Home, Hope's successor in the Chair of Materia Medica. But old habits died hard, and medicinal properties, along with anecdotes of his own use of plants in his practice at the Royal Infirmary, loom large in his elaborations on particular species. Only a handful of examples can be given here. In the garden he showed a number of temperate trees, but in the greenhouses he could show the exotics supplied by his network of overseas contacts. From the West Indies a new quinine, *Cinchona jamaicensis*,

and the cabbage tree *Geoffroea inermis* (a legume); from Africa *Brucea antidysenterica*; from the East Indies tea, ginger and the tallow tree; from North America *Myrica cerifera* from which the Natives made candles. Other economic plants included coffee, pawpaw, banana and sugar cane and dye plants such as the logwood (*Haematoxylum* – [fig. 73]). Of ornamental interest were substantial collections of cacti and mesembryanthemums, the latter showing links with South Africa, probably through Kew; and curiosities such as the Martinico rose (*Hibiscus mutabilis*) with a flower that changed from white, through carnation-coloured, to purple during the course of the day. He was also able to demonstrate the 'moving' plants *Mimosa pudica* and '*Hedysarum movens*'. There were also plants from the *Pharmacopoeia* such as squill and his particular favourite, rhubarb; and ones used for treating venereal diseases such as *Lobelia siphilitica* and *Smilax sarsaparilla* (a component of the renowned 'Lisbon diet drink' as taken by James Boswell). This brief selection will end with the 'great aloe' (*Agave americana*), of which Hope was particularly proud, and which flowered in 1767 [fig. 57]. A meteorological diary of the lead-up to its flowering was kept, and the plant drawn and engraved by Thomas Donaldson – this 'century plant' came from one of the previous gardens, and had been given (aged about 20) in 1720 to Hope's predecessor Alston by the great Hermann Boerhaave of Leyden.

History of Botany

In the spirit of a historically minded age Hope rounded off his course with a coda of two or three lectures on the history of botany. He divided this into four (un-named) periods: the first could be called Classical (Grecian, Roman, 'Times of ignorance' and Arabian); the second, the development of plant classification during the Renaissance (Brunfels to Jungius, including the neglected contributions of Gesner and Caesalpinus); the third, the beginnings of the modern period from Morison to Catesby, including Tournefort and Ray; and finally Hope's own era starting with Linnaeus, through van Royen, Rumphius, Haller, Miller, Hill, *Flora Danica* and Adanson, ending with Banks and Solander. Curiously absent from the printed abstract ('Heads' – the only such that he published) of these lectures is the name of Jussieu, but this is because Hope's approach was bibliographic, and Jussieu published nothing (from which Hope might have taken a lesson)! It is worth noting that the history of Scottish botany was not neglected: he was especially interested in Robert Morison (of whose family history he asked David Skene to make enquiry in Aberdeen), and he apparently wrote a history of the Abbey Garden at Holyroodhouse, the manuscript of which has unfortunately not been found.

BOTANICAL LECTURES IN EIGHTEENTH-CENTURY BRITAIN

It has been claimed above that Hope's course was the most advanced of its time in Britain. But what else was going on here during this period? The practical instruction of apothecaries, the chief role of early physic gardens such as Chelsea, doubtless continued, but Hope's aims and audience were much more academic. He must certainly have known, if indeed he did not attend, the pioneering botanical lectures of William Cullen in the physic garden at Glasgow in the late 1740s, and was aware of the development of the teaching of the science elsewhere in Britain – notably at Cambridge (and Oxford). He had visited Cambridge in 1766 and kept in touch with its Professor, Thomas Martyn, who had started lecturing in the 'Walkerian' garden in 1763 (Martyn charged two guineas for a first course, and one for a second). Hope owned Martyn's printed *Heads of a course of lectures in botany read at Cambridge*, published in 1764, and this syllabus makes for an interesting comparison with Hope's own. The most striking difference is its length, Martyn's consisting of only 31 lectures, but more important is the balance of the content – in Cambridge only a single lecture (the fifth, 'Of vegetation') was devoted to physiology, the rest almost entirely to terminology and classification, the last twenty a trudge through the Linnaean classes. At this time Oxford, under Humphrey Sibthorp, was moribund botanically and in 1764 Joseph Banks as an undergraduate famously had to send to Cambridge for a lecturer whom he paid out of his own pocket. The impression has sometimes been given that this was a personal tuition, but Richard Pulteney told Hope in September 1764 that 'Mr [Israel] Lyons has been reading Botanic Lectures at Oxford this summer where he has had wonderfull success having had 60 pupils in a place where Botany has not been known since Dillenius's Death'. Hope greeted this news with enthusiasm and sought further details, asking if Lyons had published 'Heads' like Martyn, 'how many lectures he gave, when he began, & if he proposes to continue?' – Hope realised that this was a new era in teaching and expressed the hope 'in a very few years to see Botany very general all over Brittain'. It is almost certain that Martyn did not use illustrations in his lectures, these not being known at Cambridge until their use by J.S. Henslow in 1827 (John Parker, pers. comm.), and it seems very doubtful that Lyons did, so one of Hope's greatest innovations, over and above the breadth and depth of the topics covered, was his outstanding use of visual aids.

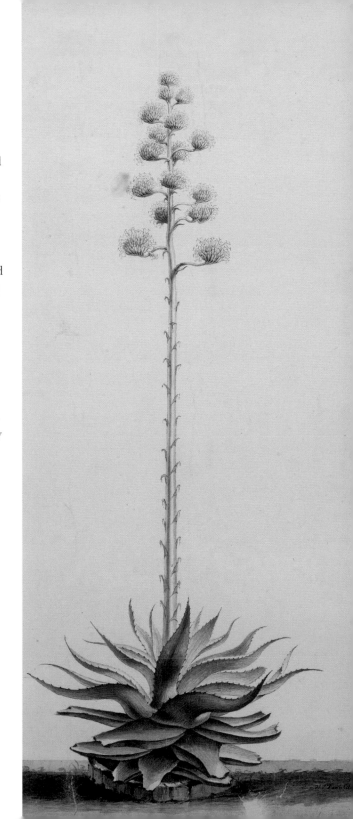

Fig. 57. *Agave americana*, 'the great aloe'. Ink wash drawing by Thomas Donaldson, 1767.

HOPE'S USE OF ILLUSTRATION

The survival of Hope's collection of teaching drawings is remarkable and unique, though even if it had been lost his use of such material would be known from the thumbnail sketches in some of the student notes. From these notes the great prominence, and legitimacy, that Hope gave to the use of taste and smell is proved, but nothing is recorded by any of the students of any similar statement Hope may have made on the value of sight. But even if not explicitly stated such a belief on Hope's part can be assumed. When in London and Oxford in 1766, he made particular comment on the botanical drawings of the great Western botanical artists Maria Sybilla Merian and Nicolas Robert, and also on the work of anonymous Chinese artists. There are also clues to the importance he placed on illustration in his historical lectures. His second period (the Renaissance) began with 'The first prints of plants by Otho Brunfelsius' – the drawings of plants from nature by Hans Weiditz published as woodcuts in Brunfels's *Herbarum vivae eicones* in 1530 are still regarded as epoch making, and Hope's copy of this great work is still in the RBGE library. In Hope's third era of botanical history the 'accurate prints of each genus' by the French botanist J.P. de Tournefort are highlighted; and many of the other authors mentioned are known for their handsomely illustrated works. The huge contribution of 'art in the service of botany' is well known, and much studied, though it is worth remembering that there were always those (Linnaeus included) who were wary of the use of the sense of sight, worried that it might lead to superficial knowledge. Clearly Hope was not among these sceptics and, while nothing specific is recorded, the very way that he taught the subject, with its strong emphasis on the demonstration of living plants and herbarium specimens, speaks for itself. From this it is not such a large step to get artists and calligraphers to make visual aids that could be seen by the whole class – by enlarging illustrations from books [fig. 42], making drawings of Hope's own experiments [fig. 41], or making large charts showing classificatory schemes [fig. 56]. But who else was using such large-scale illustrations, or was this one of Hope's most significant innovations? Much further work is required on the use of illustration in lecturing in the eighteenth century – for example, is it something that Hope might have experienced in Paris?

Interesting links and parallels are to be made between botany and anatomy, another science that relies on accurate visual identification. One of the most monumental of anatomical works, the *Tabuli sceleti et musculorum corporis humani* (1747) of the Leyden anatomist Bernard Siegfried Albinus, is illustrated with plates drawn and engraved by Jan Wandelaar, who ten years earlier had produced some of the plates (including the allegorical frontispiece) for Linnaeus's only lavishly illustrated work, the *Hortus Cliffortianus*. Albinus's plates were known in Edinburgh by Alexander Monro *primus*, even before their publication in 1747, and Monro's use of artists and engravers, and his encouragement of institutions to raise the standard of drawing in Edinburgh, have recently been studied by Joe Rock (2000). Monro employed the engraver Richard Cooper to copy the Albinus plates, and both of them were involved in 1729 in the setting up of the drawing school called the Edinburgh School of St Luke, which first operated from Edinburgh University. Cooper was a man of wide cultural interests, who set up a theatre that became a major centre of artistic activity in Edinburgh, where two artists associated with Hope worked – Jacob More and Alexander Runciman. Two other artists associated with Hope, Andrew Bell and Thomas Donaldson, were taught engraving by Cooper. The line of artists, their training, and tantalising connections with Hope continues through the Edinburgh Society for the Encouragement of Arts, Science, Manufactures and Commerce, who held meetings in the Infirmary, and the 1760 foundation of the Trustees Academy, based in the University.

The first of the Academy's drawing masters was William Delacour, who drew Hope's rhubarb [fig. 18]; a later master was Alexander Runciman under whom Andrew Fyfe won his medal for drawing in 1775, and whose firm painted Hope's lecture room. The anatomical links, or those for which evidence survives, largely concern the publication of illustrations, and Fyfe resurrected the Cooper/Monro plates after Albinus in various slightly bowdlerised forms. But one item survives that suggests that realistic illustration was used in Edinburgh anatomy teaching – a magnificent life-size engraving of a cadaver, based on a drawing by Andrew Fyfe, engraved by Thomas Donaldson who drew and engraved the Leith Walk *Agave* for Hope in 1767. This illustration dates from the tenure of Alexander Monro *secundus* in the chair of Anatomy, and from Monro *tertius's* 1840 edition of his father's *Heads of lectures on Anatomy*, the middle Monro (Hope's contemporary) was said to have derived great benefit from the use of 'mathematical calculations or diagrams, to illustrate the effect of compound muscular action' (Matthew Eddy, pers. comm.). Such geometric diagrams seem quite different to anything that Hope used, but it is entirely possible that there may once have been a corpus of anatomical teaching drawings (both realistic and geometric) made for the Monro dynasty that has not survived.

The Hope collection of teaching drawings

Given the importance of drawings in Hope's teaching practice, and the extensive use of his collection in illustrating this book, something must be said of them. They fall into various categories, the largest being the teaching drawings made by the 'gardener-assistants', especially Andrew Fyfe, John Lindsay and John Bell. Agnes Williamson, the head gardener's daughter, also made some – an early example of a woman making scientific drawings that will surely be of interest to feminist art historians. Many of these drawings were enlarged (and variously modified) copies from book illustrations, especially from Hales's *Vegetables staticks* of 1727, and from the anatomical works of Sir John Hill, Marcello Malpighi and Nehemiah Grew. But many of the drawings are original and, while highly stylised, record observations or experiments made from life in the Leith Walk garden – the development of bud scales, physiological experiments on transpiration (by means of ligatures or incisions) and growth (by insertion of lead plates), pollen, sleep movements of leaves, and experiments on the effects of light and gravity. As pointed out by Francis Darwin, who was shown some of these drawings by Bayley Balfour and in 1909 wrote a paper on experimental aspects depicted, many drawings exist in two versions – an initial sketch in red chalk, and a worked-up one in grey ink wash. The other major

Fig 58. *Pandanus leram*, Carnicobar breadfruit. Probably by Gavin Hamilton, c. 1780. Ink wash.

Fig. 59. *Bletilla striata*, hyacinth orchid. Anonymous Chinese artist, late eighteenth century. Body colour and gum arabic.
This may be from Hope's collection of Chinese drawings.

category of teaching material comprises large charts showing classificatory keys in beautiful copperplate calligraphy: only these show evidence of the way they were displayed (suspended by cloth loops) [fig. 56], leaving one to wonder if the other drawings were handed round or displayed on an easel.

Hope's unpublished papers show the great value he placed on his teaching drawings. There is a listing of the groups by topic, which suggests that remarkably few have been lost as these correspond with the annotated covers in which they were originally kept, which, remarkably, have also survived. The notes also show the care Hope took of the drawings, details of the chests in which they were stored, and how they were moved from one place to another after use in a particular year's lectures. Some of the drawings are even annotated with notes on the year in which they were or were not used, giving an extraordinary sense of closeness to Hope's teaching practice.

The other material in Hope's library of visual material is not so explicitly didactic, but was probably shown at least to the keener of the students – drawings of interesting plants sent to him by former pupils such as John Lindsay from Jamaica [fig. 13], a drawing of *Pandanus leram* sent by Gavin Hamilton

from the Nicobar Islands [fig. 58], and the *burrum chundalli* from James Kerr in Bengal [fig. 14]; and from closer to home, Norfolk, *Holosteum umbellatum* sent by Charles Bryant, a friend of J.E. Smith. Hope also had a volume of Chinese drawings (two at RBGE just might be survivors from this collection of eighteenth century export art) [fig. 59]. Also included are rare proof engravings, including unfolded ones from the *Philosophical transactions*, and some that appear never to have been issued, such as the plates of various *Cinchona* species after drawings by John Hawkins [fig. 60], which must have been sent to Hope by one of his London correspondents. Last are the fine botanical drawings commissioned by Hope – John Lindsay's *Buddleja* [fig. 17]; William Delacour's *Rheum* [fig. 18]; Thomas Donaldson's *Agave* [fig. 57]; those by James Robertson of which only the *Ruppia* survives [fig. 26], and a handful of other more sketchy drawings of plants clearly made in the Leith Walk garden and probably the work of Andrew Fyfe and/or Robertson.

Fig 60. *Cinchona officinalis*, quinine. Engraving by James Mynde (with ink manuscript additions), after a drawing by John Hawkins (Hawkeens) made in 1739. Published 1741.
Hope lent this print to Richard Pulteney who had the left-hand part (which is squared for copying) re-engraved for his 1764 MD thesis on quinine.

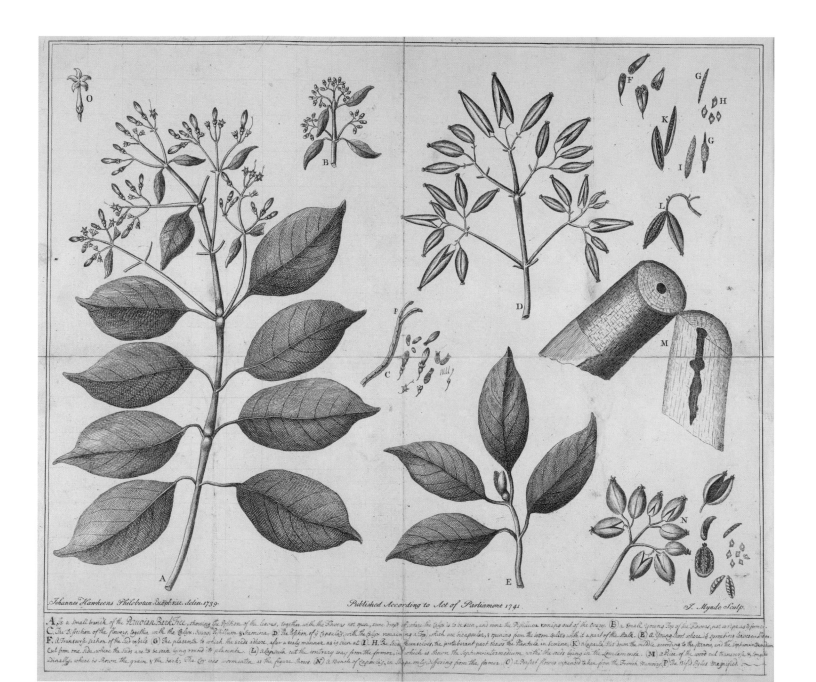

Hope's students

A physician who attended the madness of King George III [fig. 63]; a baron of the Russian Empire who, at twenty, had helped his father inoculate Catherine the Great against smallpox; the man who 'bowdlerised' Shakespeare; the surgeon on the Macartney mission to China; both father and uncle of Charles Darwin; two revolutionary Irish nationalists; a signatory to the American Declaration of Independence; a breeder of merino sheep who figures in the letters of Jane Austen ...

These are merely a small selection of some of the more remarkable characters among the cast of more than 1700 men who attended Hope's lectures in Botany (1761 to 1786) and Materia Medica (1761 to 1767). That these are known is due largely to Hope's own meticulously kept student class lists. There are certain problems both with these, and with the contemporary Edinburgh University matriculation records, due to variations in spelling; the absence of Christian names for some, and, when these are given, the severe lack of imagination on the part of Scottish parents in their choice (and parsimony in rationing these to one). Nevertheless, it is possible to follow up much about the lives and careers of many of these students, though only a start has so far

been made. Some rough indication of the eminence achieved by these products of the Edinburgh medical school is shown by the fact that, of those on Hope's lists, at least 66 merit an entry in the *Oxford dictionary of national biography* (in similar works for America and Canada, nine and one respectively); 21 went on to hold academic chairs in Scotland,

Fig. 61. Botany class ticket, given to Stephen Pellet for the year 1777. Unknown Edinburgh engraver. Edinburgh University Library, Centre for Research Collections.

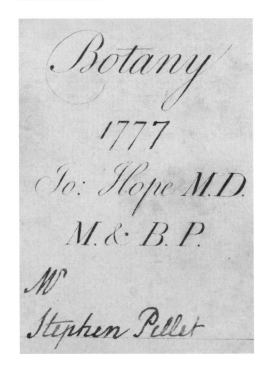

England, Ireland and the United States; and more than twenty became Fellows of the Royal Society.

The reason for Hope's records is, at least in part, his somewhat mercenary nature – they were primarily kept to show who had paid for his annual class ticket (which cost 2s 6d) [fig. 61] and lecture fee (2 guineas, later 3), and, more particularly, those who had not (the Irish were particularly often offenders). But the lectures must have been enjoyable as many individuals attended two, three or even more times, for which occasionally Hope gave a reduction. Most students belonged to the medical faculty, but the timing of the lectures was unpopular as, to suit the growing season that allowed of plants for demonstration, they took place between May and July, long after the other lectures had finished. Nonetheless Botany was a compulsory part of the course and all students who wanted to graduate (by no means all) had to attend at some point during their period of study – most commonly three years, but often considerably longer. For reasons that are not apparent numbers at the Botany lectures fluctuated considerably – averaging 71, but reaching a peak of 102 in 1775 and a nadir of 27 in 1765.

It is perhaps necessary to make a distinction between students who simply attended the lectures and those who could in some sense be considered 'pupils'.

Fig. 62. William Withering (1741–1799). Materia Medica student, class of 1763. Engraving by William Bond, after a portrait by Carl Frederik von Breda (published 1822). In his left hand Withering holds a spike of foxglove.

Fig. 63. Sir Lucas Pepys (1742–1830). Botany student, class of 1765. Stipple engraving by James Godby, after a portrait by Henry Edridge (published 1809). Royal College of Physicians of Edinburgh.

Of the former it is impossible to tell the influence of Hope in their subsequent, usually medical, careers. These took many forms – physicians in towns throughout Britain and Ireland, and in the army and the navy. Several treated royalty, but many more were at the forefront of the medical reforms of the second half of the eighteenth century, in vaccination against smallpox, in setting up county hospitals, in the more humane treatment of the insane, and in the beginnings of public health by means of statistical analysis of incidence of disease, and records of births and deaths. The students who might be considered pupils are those who went on to contribute to botany, or related fields such as agriculture. Another category of student was that whom Hope did not charge fees – these included fellow academics and their offspring, and other professionals or landowners with improvement or botanical interests. Some of the most important of these pupils (Buchanan, Lindsay, Menzies, Roxburgh, J.E. Smith) have already been discussed under relevant headings, but some significant others will be discussed below.

GEOGRAPHICAL ORIGINS

The tenure of Hope's botanical posts was at the height of the Enlightenment, and the peak of renown of the Edinburgh medical school (centred, especially, around William Cullen), which drew students to the 'seat of the medical muse' from far and wide. As the flow of information was not only from teacher to pupil – in his lectures Hope often quoted observations gleaned from students – it is worth being a little more specific about the extraordinarily wide geographical origins of those who attended his classes. From the New World 77 came from North America (eight states, with the greatest numbers from Pennsylvania, Virginia and the Carolinas) and 35 from the West Indies (especially St Kitts and Antigua). From Europe: five from Switzerland; three from Russia; two from each of Germany, Denmark, Sweden, Holland and France; and one each from Italy, Spain and Portugal. For natives, the precise place of origin is seldom specified, the majority being denoted merely British. Although the majority of these must have been from Scotland, there must surely have been more than the 121 specifically noted as from England (and four from Wales). Of the English contingent a large number, given religious bars of access to Oxford and Cambridge, came from dissenting backgrounds, and several had previously attended the Warrington Academy and

must have been pleased to hear Hope's account of the work of Joseph Priestley. Those from across the Irish Sea were identified as such, and of these 120 attended Hope's lectures.

BOTANISTS AND NATURALISTS

Most of Hope's pupils who distinguished themselves in the field of botany have already been discussed, but there are several others who, through their botanical publications, contributed to the spread and popularisation of the science.

The first of these, and among his earliest pupils, is William Smellie (1740–1795) (EP19), a considerable figure of Enlightenment Edinburgh whose neglect is surely due, at least in part, to his unfortunate surname. He attended Hope's Botany lectures in 1762 and 1763 while working as a printer's corrector for Murray & Cochrane, who, with premises in Craig's Close, were Hope's neighbours. It is reported that when Hope sprained his leg badly he was so impressed with Smellie's knowledge that he asked him temporarily to take over the lectures, which he did 'to the entire satisfaction of his fellow students'. In 1764, with the help of his reading boy Pillans, Smellie collected a herbarium of 400 dried plants, which he presented to Hope who thought it worth 'some honorary reward' and that

Smellie might win the gold medal the following year (he did not, Adam Freer got it). It was in this same year, 1764, that Andrew Duncan attended Hope's class and reported that 'at that time our worthy professor ... encouraged the most ingenious and able of his students to read essays to the class on botanical subjects. These were stiled [after Linnaeus] the Amoenitates Academicae; and one of the first I heard was an Essay by [Smellie] on the Sexes of Plants'. It was claimed by Smellie's biographer (Robert Kerr, another Hope student) that this essay received a gold medal, but this is probably incorrect. That it argued against sex in plants, and went against Linnaean orthodoxy, mattered not to Hope as it was well argued, and Smellie published it both as a pamphlet at the time and later in his *Natural history of philosophy*. The essay was also published in one of the enterprises for which Smellie is best remembered – the first edition of the *Encyclopaedia Britannica*. This great work was the brainchild of Andrew Bell (EP18), its principal proprietor whose fortune it made, a fine engraver who had made the plates of *Rheum* and *Eriocaulon* for Hope. For £200 Bell asked Smellie to write accounts of 'fifteen capital sciences' for the *Encyclopaedia*, which included that on Botany in which tribute is paid to Hope. Smellie was disarmingly frank about this work – making 'a Dictionary of Arts and Sciences with a *pair of scissors*, clipping

Fig. 64. John Sims (1749–1831). Botany student, class of 1772. Etching by Maria Dawson Turner, after a medallion by Benedetto Pistrucci (1817). Director and Board of Trustees, Royal Botanic Gardens Kew.

Fig. 65. John Berkenhout (1726–1791). Botany student, class of 1763. Engraving by Thomas Holloway, after his own drawing (published in the *European magazine* 1788). Royal College of Physicians of Edinburgh.

linguae botanicae Linnaei, a useful English glossary of Linnaean terms, published in 1764 and dedicated to Hope – this work shows the early date from which Hope was teaching Linnaeus's natural orders. Another botanical author, Arthur Broughton (c. 1758–1796) from Bristol, studied with Hope in 1777; he returned as a physician to the Royal Infirmary in his home town where, in 1782, he published *Enchiridion botanicum*, with brief diagnostic descriptions ('essential characters') in Latin of the genera and species of British plants. Broughton's health gave way and he went to Jamaica, where he published several editions of *Hortus Eastensis*, a catalogue of the plants in the garden of Hinton East Esq.

Although William Withering (1741–1799) attended Hope's 1763 Materia Medica classes, he conspicuously did not attend the botanical ones [fig. 62]. As his biographer wrote:

out from various books a *quantum sufficit* of matter for the printer' – and one might speculate that one of the 'books' he cannibalised was his own notes from Hope's lectures. Hope's financial aid in setting Smellie up in his own printing business in 1765 has already been noted, but this is not the place to elaborate further on his interesting subsequent career, which involved an appointment as keeper of the museum of the Society of Antiquaries, and his distinguished translation of the Comte de Buffon's *Histoire naturelle*.

As third editor of the *Botanical magazine* John Sims (1749–1831) [fig. 64] is, after J.E. Smith, perhaps the most important of Hope's pupils in the realm of taxonomic and horticultural publication. Sims succeeded the founder William Curtis and his brother Thomas and edited the 'Bot. Mag.' from 1801 to 1826. John Berkenhout (1726–91), born in Leeds, had led a rather adventurous life as a captain in the Prussian infantry, then a British regiment, before coming to Edinburgh and taking Hope's class in 1763 [fig. 65]. As a mature undergraduate, with some assistance from Arthur Lee, he compiled a *Clavis Anglica*

in Mr Withering we have a remarkable instance of the versatility of taste, or rather, perhaps, of the influence of prejudice. He who afterwards became the author of a work not inaptly described as 'the most elaborate and complete National Flora of which any country can boast', has this expression in a letter to his parents:– 'The Botanical Professor gives annually a gold medal to such of his pupils as are most industrious in that branch of science. An incitement of this kind is often productive of the greatest emulation in young minds, though, I confess, it will hardly have charm enough to banish the disagreeable ideas I have formed of the study of botany'.

The reference to a 'National Flora' is to Withering's *Botanical arrangement of all the vegetables in Great Britain,* of which he sent Hope a copy of the first edition in 1776. By this time Withering was working as a physician in Birmingham, and was a member of the Lunar Society and friend of Erasmus Darwin. His best known work was on the medicinal use of the foxglove, and he duly sent Hope a copy of his 1785 publication on the subject. The *Botanical arrangement* was a huge success and ran to fourteen editions, for the second of which Withering was helped by Jonathan Stokes who had attended Hope's lectures in 1780. The work was unillustrated so Stokes's contribution was to add 'a new set of references to figures', drawn from the whole of botanical literature, citing them 'in order of supposed comparative excellence'. This ended acrimoniously, and Withering had to go to law to reclaim the numerous books Stokes had borrowed in order to undertake this enormous task. Another medico-botanical writer was the Quaker William Woodville (1752–1805), who took Hope's Botany class in 1773 and 1774 [fig. 67]. After an unfortunate incident in Cumberland, when Woodville accidentally shot and killed an intruder in his garden, he went to London where he became active in the movement to prevent smallpox by inoculation and, later, by vaccination. It was while at the Smallpox Hospital at St Pancras, where he had a small botanic garden, that Woodville published his four-volume *Medical botany* (1790–5), in which all the species are illustrated by plates drawn and engraved by James Sowerby.

The last 'natural philosopher' is mentioned because he prefigures RBGE's twentieth-century Chinese interests. Hugh Gillan (d. 1798) studied with Hope in 1784, and was recommended by Sir Joseph Banks as surgeon on the Macartney Embassy to China of 1792–4. Gillan was not the botanist on the mission, and there is no record of his collecting any plants, but at the end of the trip he took the botanical specimens collected by Sir George Staunton and David Stronach from the ship to Macartney's house, and examined the state of the living tea plants they brought back. He is remembered for his scathing 'observations on the state of medicine, surgery and chemistry in China' – described as a 'tangle of misstatements', though this was not entirely his fault as the Chinese had been strictly forbidden to provide the barbarian foreigners with any information.

Fig. 67. William Woodville (1752–1805). Botany student, class of 1773 and 1774. Stipple engraving by William Bond, after a painting by Lemuel Francis Abbott (published by R.J. Thornton, 1806). The botanical 'emblem' above the portrait is hemlock. Royal College of Physicians of Edinburgh.

Fig. 68. Daniel Rutherford (1749–1819). Botany student, class of 1764, 1765 and 1767. Stipple engraving by William Holl, after a painting by Henry Raeburn (published by R.J. Thornton, 1804). The botanical 'emblem' above the portrait is (for reasons unknown) *Nicandra physalodes*.

ACADEMICS

It was noted above that 21 of those attending Hope's lectures went on to become professors. The influence of Hope on most of these (for example Smithson Tennant, Professor of Chemistry at Cambridge; Edmund Cullen, of Materia Medica at Dublin; and Robert Perceval, of Chemistry, also at Dublin) was probably slight, but John Parsons, at one time Professor of Anatomy at Oxford, and winner of Hope's fourth gold medal, assisted Lightfoot with his *Flora Scotica*. It is, however, worth saying a little more about the students who held Scottish chairs. In the filling of these posts there was a notorious hereditary tendency at work, but though Thomas Charles Hope wanted to succeed his father this did not happen, and through Henry Dundas, who had substantial influence over the appointment, the Botany chait went to another former pupil – Daniel Rutherford. All was not lost and the younger Hope, after a sojourn with his maternal uncle in Glasgow, did later succeed to an Edinburgh chair (Chemistry).

Daniel Rutherford (1749–1819) was the son of a prominent Edinburgh physician, and an uncle of Sir Walter Scott [fig. 68]. That his botanical interests were serious is shown by the fact that he took Hope's Botany class three times (1764, 1765, 1767). His MD thesis, on fixed or mephitic air (carbon dioxide), however, revealed his great interest in and talent for chemistry. This topic, as relayed to his students by Hope, had previously been studied by Joseph Priestley, but Rutherford probably independently showed that when air was respired or passed over burning lime, in addition to the fixed air there was a residual component later known as nitrogen. After graduating, Rutherford went on a tour of France and Italy, and took time on the slopes of Mount Etna in Sicily to measure for Hope an object of great interest to him, an ancient chestnut, the 'Castagno dei Cento Cavalli' [fig. 36]. Despite chemical interests and duties as a physician at the Infirmary, and despite claims to the contrary, Rutherford did take his botanical teaching seriously at least in the earlier of his 33 years in office, as may be seen from surviving summaries of his lectures. By means of a series of excellent head gardeners Rutherford was able to continue Hope's tradition of demonstrating a wide range of plant material, which, with his teaching of natural classification, was a major influence on Robert Brown, Britain's greatest botanist of the nineteenth century.

In 1761 an Edinburgh physician, Robert Ramsay, attended Hope's Botany class. An enigmatic figure, whom Hope regarded as 'a particular friend', it was Ramsay who delivered his mentor's letter to Bernard de Jussieu in 1765, while passing through Paris on the way to Nice, where he was to spend the winter as physician to the Earl of Breadalbane, where Hope asked him to collect Mediterranean algae for John Ellis.

In 1767 Ramsay was appointed Regius Keeper of the Edinburgh University Museum and Regius Professor of Natural History, though the Town Council only ratified these appointments in 1770. The once great natural history collections of Balfour and Sibbald had probably decayed before Ramsay's appointment, but even so he was notably inactive in these posts. He died in 1778 and was succeeded by the Rev John Walker, an altogether more active and effective natural historian.

In providing a fulsome obituary of his teacher, posterity owes Andrew Duncan (1744–1828) (EP10) a considerable debt, since otherwise many details of Hope's life would have been left unrecorded. Duncan attended the Botany class in 1764, and must have made an impression, as Hope asked him to collect seaweeds for John Ellis while 'settled in a Gentleman's family in [his native county of] Fife' in the summer of 1766. Hope's and Duncan's lives coincided at many points thereafter, not least in the clubs and societies to which both belonged, several of which the genial Duncan established to build bridges between antipathetic bodies of surgeons and physicians. John Gregory, Professor of the Institutes of Medicine, died in 1773, and Alexander Munro Drummond (who attended Hope's lectures in 1769) was appointed his replacement, but declined to take up the position – life as physician to the King of Naples suiting him better. While waiting for Drummond's decision Francis Home

THE AMERICANS

The students from the American colonies who attended Hope's lectures form a fascinating group, but as their huge influence in the emerging United States lay especially in the fields of medical education, in politics and diplomacy, and in the abolition of slavery, only a handful of the most important of them can be mentioned here. Arthur Lee (1740–1792), an Old Etonian of a wealthy family with large estates in Virginia, won the first of Hope's gold medals in 1763, became well known as a polemicist and diplomat, a friend of Benjamin Franklin, and for a time American ambassador to France. The winner of the second medal, Samuel Bard (1742–1821) [fig. 69], takes us north to New York; he held Chairs in the Theory and Practice of Medicine, and in Natural Philosophy, at what became Columbia University. All of Bard's generation were caught up in the revolutionary wars and, despite loyalist sympathies, while George Washington's government was based in New York, Bard was the general's private physician. Also at Columbia was Samuel Latham Mitchill (1764–1831), remembered as a naturalist, physician and legislator. From 1792 to 1801 he was Professor of Botany, Natural History and Chemistry at Columbia; he founded the first medical journal in the United States, served in the House of Representatives, and for a time was Senator for New York.

The remaining American students to be mentioned were all based in Philadelphia and its college, which became the University of Pennsylvania. Benjamin Rush (1746–1813) was Professor of Chemistry and later of the Theory and Practice of Medicine. Having been a surgeon in the Continental Army, he is considered a founding father of the United States, being one of the signatories of the Declaration of Independence. Later he was known as a rather old-fashioned, phlebotomising (blood-letting) physician, but also as a keen abolitionist. William Shippen (1736–1808) was also a surgeon with the revolutionary army, became Professor of Anatomy, Surgery and Midwifery at the College of Philadelphia, and one of the founders of the American Philosophical Society. Caspar Wistar (1761–1818) [fig. 70], another abolitionist, succeeded to Benjamin Rush's Chair in Chemistry, and was also Professor of the Institutes of Medicine, and later of Anatomy, Midwifery and Surgery, but his name is best known to gardeners and botanists through Nutall's (misspelt) generic name *Wisteria*.

Fig. 69. Samuel Bard (1742–1821). Botany student and gold medal winner, class of 1763 and 1764. Aquatint and engraving by W. Main, after a painting by John Vanderlyn (published by J. Thacher, 1828). Royal College of Physicians of Edinburgh.

Fig. 70. Caspar Wistar (1761–1818). Botany student, class of 1785. Lithograph by W.S. & J.B. Pendleton, after a painting by Bass Otis (published by J. Thacher, 1828). Royal College of Physicians of Edinburgh.

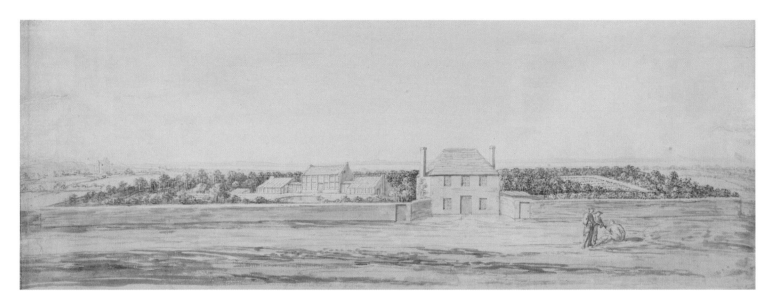

Fig. 71. Perspective view of the Leith Walk garden. Watercolour by Jacob More, 1771.

This drawing shows the unique teaching and research institution familiar to Hope's students. Lectures took place in the upper floor of the gardener's cottage; demonstration of living plants in the greenhouse and hothouses and the main part of the garden; of medicinal plants in the 'Schola Botanica' seen above the heads of the group of figures. The latter is a symbolic representation of the healing arts, as undertaken by the majority of Hope's students in their subsequent medical careers.

Hope's legacy

It is appropriate to end with a few words summarising Hope's strengths, his achievements and legacy, and raise some questions about reasons that might account for why his influence was not greater, and why he has been largely forgotten – leading to pointers to work still required.

To start with some possible reasons for his lack of influence. The first of these, without doubt, is Hope's lack of publications. No matter how many students are taught, in a science – particularly botany, which is so heavily bibliographical and historical – not to publish is to consign oneself to oblivion. It means that influence is to a great extent restricted to the students who attended the lectures. Given the great number of these (over 1700), the direct influence, while hard to measure, seems disappointingly slight.

Another reason, which cannot be laid at Hope's door, results from the professional lives his students went on to lead – almost all were medics who had to make a living as physicians and surgeons. Despite Hope's medals to encourage collecting, two won by Americans, neither of these, nor any of the numerous other American students contributed anything towards the botanical exploration of their vast continent when they got back home, probably because they were simply too busy doctoring and teaching. It is significant that Hope's most prolific pupil in terms of publications was J.E. Smith, who had ample private means. The limiting effect of this yoking of botany to medicine will be returned to. But there was one source of medical employment that did officially allow of some botanical exploration and development work – the patronage of the Honourable East India Company. Although for the Company botany was always an activity subsidiary to economic and commercial goals, it allowed the major contributions of two of Hope's most significant pupils. However, either in medical practice, or Company employment, it was hard to do more 'academic' botanical work in a colonial context – the experimental was virtually an impossibility, and even microscopic work was extremely difficult (making John Lindsay's work on the germination of fern spores, and his experimental work on the sensitive plant in Jamaica the more remarkable), so botanical work was limited largely to collecting, recording and classifying. Of such cataloguing and documentary work Hope's influence was direct and important both in Scotland, through James Robertson, and in India, with William Roxburgh and Francis Buchanan (though the effectiveness and memory of both first and last of these was strictly limited, as neither published their floristic work).

The second most important direct influence of Hope was arguably in the use of illustration in the pictorial documentation of flora and materia medica – in this the contributions of Roxburgh in India [fig. 72], and J.E. Smith and William Woodville in the United Kingdom [figs 73, 74], were outstanding.

Influence is not entirely restricted to those taught personally, but passes on in an apostolic succession: two pupils continued to teach along Hopean lines – Daniel Rutherford in Edinburgh (for more than 30 years) and William Hamilton in Glasgow (for only nine). Little is known about Hamilton or his pupils, but his classes must have been small; it is through Rutherford that the most important secondary influence comes – for he taught Robert Brown, Jupiter Botanicus. By this time Jussieu's work had been published, and the huge influence of Brown's promotion of the natural system is well known, and can be linked, through Rutherford, back to Hope's tentative steps.

Fig. 72. *Indigofera atropurpurea*. Unknown Indian artist, Calcutta, c. 1805. Watercolour and gum arabic.
Drawn for William Roxburgh in the Calcutta Botanic Garden from a specimen sent by Francis Buchanan from his expedition to Nepal (the first such by a Western botanist) in 1802.

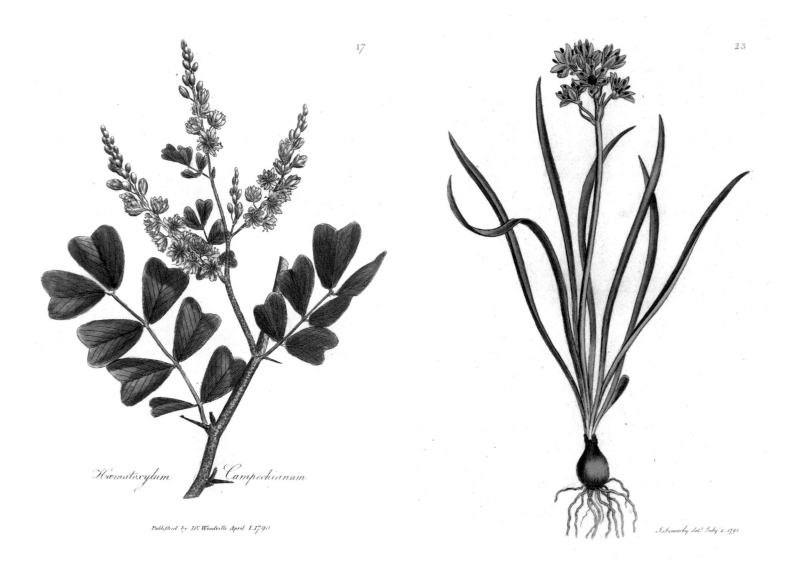

17

23

Hæmatoxylum Campechianum

Published by Dr Woodville April 1.1790.

J.Sowerby del.t July 1. 1791.

Fig 73. *Haematoxylum campechianum*, logwood. Hand coloured engraving by James Sowerby, published in William Woodville's *Medical Botany* (1790).
A West Indian plant shown to students at the Leith Walk garden.

Fig 74. *Scilla verna*, spring squill. Hand coloured engraving by James Sowerby, published in J.E. Smith's *English botany* (1791).
First recorded in Britain by James Robertson in Sutherland in 1767.

What were Hope's achievements? Hope was above all a great synthesiser, and the areas over which he cast his gaze were, for his time, unusual in their breadth. The knowledge synthesised was acquired in various ways – of these the literary, through reading, has always been (over?) emphasised, and it is easy to forget (because it leaves little trace) how important was the activity of listening – in Enlightenment times as much as in our own. The role of the Edinburgh debating societies has been noted in Chapter 1, but Hope was also listening to more vernacular sources, as graphically displayed on the cover of this book. A great deal can be discovered of Hope the reader from the books in his library and the references cited in his lectures. In addition to scholarly works these covered what might be termed semi-popular literature, such as the *Gentleman's magazine* and published 'Travels'.

The scholarly works included the field of medicine, but this aspect of Hope's work is unstudied and largely unknown. He undoubtedly read widely in the subject and had relevant books in his library, but other than those relating to materia medica these have largely not survived – despite shedding his chair in the subject Hope's interest in materia medica was lifelong, and certainly formed part of his synthesis. His knowledge of classification encompassed the whole of the existing literature, and this was one

area where he did undertake original work: his work on a natural classification, but this was not resolved, let alone published. Work on classification leads to Linnaeus, and from him Hope took much more than classification – notably the more applied aspects of Linnaeus's utilitarian agenda, which led into areas that later developed into ecology and ethnobotany. To these topics (and far more so than his nearest contemporary Thomas Martyn) Hope added the anatomical work of Malpighi and Grew, the great physiological work of Hales, and the French school of Duhamel. Hope's ability to read French was important and in his notes and lectures he made as frequent reference to the Parisian *Mémoires de l'académie royale des sciences*, as to the most recent volumes of the London-published *Philosophical transactions of the Royal Society*. The contemporary issues that Hope failed to pick up on are extraordinarily few – given his interest in French scientific literature and the fact that he taught two sons of Erasmus Darwin, it is curious that the question of mutability of species is not mentioned once (at least through the distorting lens of the student notes); and given his 'improvement' agenda it is perhaps surprising that there is next to nothing about plant breeding (a necessary precursor to genetics) – Hope's interest in hybridisation, like that of Linnaeus, seems to have been more for the taxonomic complications it led to.

It cannot be claimed that Hope was a very 'original' scientist (and what was previously believed to be the one exception, the experiments on the competing effects of light and gravity on growth, proves not to have been his). But the question of originality is debatable: the craving for it, and the award of status to those deemed to have achieved it, in our celebrity-driven culture is well known. In the arts this doubtless has origins in romantic ideas of genius, and in the history of science, especially those written by scientists, an approach that could be characterised as 'winner takes all' – an interest in those who 'got it "right"' and led to 'progress', and consignment to oblivion for the rest. It is interesting to note that as early as 1909 Francis Darwin in his sympathetic account of Hope the physiologist was infected with this approach, thinking it worthwhile only to discuss the experiments that most closely approached the 'original'.

But it must not be forgotten that the confirmation or, as importantly, refutation, of the experimental results of others is one of the major means by which knowledge accumulates and is consolidated (though only if one publishes the results!) – in this field Hope was exceptional. It was not enough for him to read the experiments of Hales and Duhamel; he felt compelled to redo them (with slight modifications) at Leith Walk on his beloved trees [fig. 75]. This takes us back to what, with the benefit of hindsight, appear to be limitations of the time in which he lived. Botany in the eighteenth century was in its infancy; it was still largely an applied science, wedded to medicine (materia medica) and the improvement agenda of boosting commerce and agriculture. The great steps forward came only later, when it was linked with the physical sciences of chemistry and physics, including the technology that allowed of more accurate measurement and analysis. In Hope's day, due to limitations in microscopy, observation was limited largely to visible phenomena measurable by relatively crude devices. There was no way Hope could understand the fundamentals of 'absorption' without knowing of physico-chemical processes including osmosis, or 'perspiration' without even knowing of stomata (which he never mentioned under any name). Hope was aware of the importance of chemistry and reported the experiments of Joseph Priestley and Thomas Percival, but did not have sufficient interest to repeat these himself, though his pupil Daniel Rutherford did.

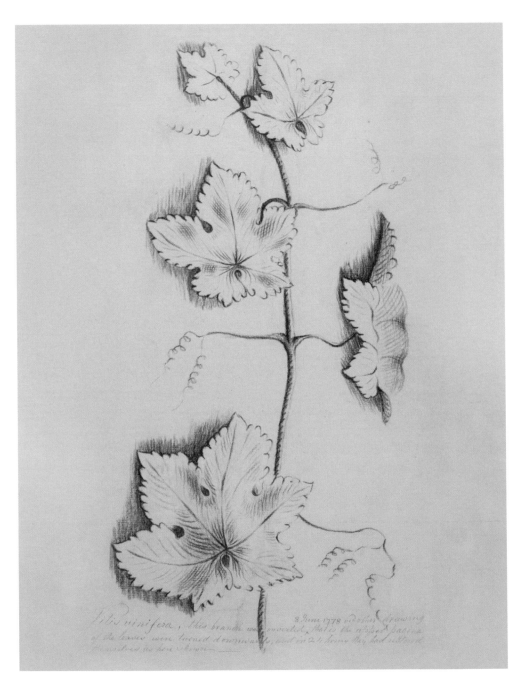

Fig. 75. Hope's experiment (repeating one first carried out by Charles Bonnet) on the restoration of leaf position in an inverted branch of grape vine.

On 8 June 1778 Hope inverted the branch so that the upper leaf surfaces were facing downwards (left); in 24 hours their correct orientation had been restored (right).

Given the period in which he lived, the life and work of John Hope can be seen to have been a remarkable one, and worthy of celebrating and examining in depth. The purpose of this publication has been to put a little more flesh on the bones so ably drawn by Alan Morton in 1986, but much further work remains to be done to understand Hope's work, both in its own right, and in the context of his time, to re-establish him in his rightful place at the centre of the Edinburgh Enlightenment. The priorities are a critical edition of Hope's lectures, a detailed study of the Leith Walk garden, and a study of Hope's medical practice. The last has been completely ignored in both editions of this book, but given the time and effort Hope devoted to its pursuit, it needs to be considered for a complete understanding of Hope's thought, as it surely informed and influenced his approach to botanical science.

APPENDIX 1.
A bibliography of Hope's publications

Hope's publications are few in number, though slightly more numerous than previously realised. They may be categorised as follows: an edition of the lectures of his late teacher; four scientific papers (plus a description contributed to a Linnaean work); two garden catalogues; and five student texts (some in two editions).

The reasons for this paucity are various, though it is worth noting that the same applies to several of his scientific (if not his more literary) contemporaries – Joseph Black was similarly reluctant to publish and William Cullen did so only later in life, both of these being spurred into print by the publication of pirated editions of student notes. One reason was Hope's (self-imposed) need to concentrate on teaching to boost student numbers to maximise income from student fees – and more than half of Hope's works (if one includes the garden catalogues) can be seen as teaching aids. But lectures only occupied three months, and even with generous preparation time, this left a very substantial part of the year for writing, had he been so minded. Hope was undoubtedly keen on money and explicitly stated in a letter to Pulteney that while in the prime of life it made more sense to build up a private medical practice than to publish. But there was probably more to it than this, and that he had a genuine unwillingness to commit his thoughts to print. The papers on *Rheum*, *Eriocaulon*, asafoetida, and especially that on *Buddleja*, are slight in the extreme. That he did not produce an edition of his lectures (or even 'Heads' for them) suggests a diffidence confirmed in his personal papers – that he was always striving to improve their content and arrangement, and that he was never fully happy with

them. It should also be remembered that he died unexpectedly, and, had he lived longer, might perhaps like Cullen have prepared his lectures for a publisher. If not the lectures as a whole, he might have worked at least his ideas on a scheme of natural classification into publishable form.

HOPE, J. (1766). Extract of a letter from Dr. John Hope, Professor of Medicine and Botany in the University of Edinburgh to Dr. Pringle [on *Rheum palmatum*]. *Philosophical transactions of the Royal Society* 55: 290–3, tt 12, 13.

HOPE, J. (ed.) (1770). *Lectures on the materia medica: containing the natural history of drugs, their virtues and doses: also directions for the study of materia medica; and an appendix on the method of prescribing. Published from the manuscript of the late Dr. Charles Alston ... by John Hope, M.D. ... 2 vols.* London: printed for Edward and Charles Dilly ... and A. Kincaid and J. Bell, at Edinburgh. 4to.

HOPE, J. (1770). A letter from John Hope, M.D. F.R.S. Professor of Physic and Botany in the University of Edinburgh, to William Watson, M.D. F.R.S. on a rare plant [*Eriocaulon*] found in the Isle of Skye. *Philosophical transactions of the Royal Society* 59: 241–6, t. XII [A. Bell, after I. Robertson].

ANON. [HOPE. J.] (1770). *Termini botanici: in usum juventutis academicae Edinensis. Accedunt. index, rerumque series.* Edinburgi: Apud Balfour et Smellie. 8vo.

ANON. [HOPE, J.] [undated]. [Untitled – informally 'Systems']. [Edinburgh]. Five printed sheets, c. 250 x 205 mm, giving summaries in the form of keys of seven historical natural classification systems.
[1] Methodus A. Cæsalpini, Methodus A.Q. Rivini;
[2] Methodus Calycina C. Linnæi;
[3] Clavis Classium Raii; [4] Clavis Classium A. Van Royen; [5] Methodus D. I.P. de Tournefort, Methodus D. F.B. Sauvages, ex foliis.

ANON. [HOPE, J.] (1771). *Genera plantarum, ex editione duodecima Systematis naturæ, illustrissim. Carol. a Linne. In usus academicos.* Edinburgi. Typis Academicis. Pp [i–ii], tab. 'Clavis Classium', [1–]3–88. 8vo.

HOPE, J. (1771). Eriocaulon (characteres reformati), in *Mantissa plantarum altera* p. 167, C. Linnaeus. Holmiae: impensis direct. Laurentii Salvii.

ANON. [HOPE, J.] (1775). *A catalogue of trees and shrubs growing in the botanic garden at Edinburgh.* Edinburgh [no publisher given]. Pp [1–]4–17. 8vo.

ANON. [HOPE. J.] (1778). *Termini botanici: in usum juventutis Academicae Edinensis. Accedunt index, rerumque series.* Edinburgi: Apud Balfour et Smellie. Pp [1–]3–39, and tt I–V. 8vo.

Note. Similar to the first edition, but lacking the preface and classification schemes.

ANON. [HOPE, J.] [undated]. [Untitled – informally 'Systems']. [Edinburgh].

Apparently a second edition of Hopes' 'Systems'. Four printed sheets, c. 250 x 205 mm, giving summaries in the form of keys of five historical natural classification systems. [1] Methodus D. I.P. de Tournefort; [2] Methodus A. Cæsalpini, Methodus A.Q. Rivini; [3] Clavis Classium A. Van Royen; [4] Clavis Classium Raii.

ANON. [HOPE, J.] (1778). *Catalogus arborum et fruticum in horto Edinensi crescentium anno 1778.* Edinburgi: Apud Balfour et Smellie. Pp [1–]4–20]. 8vo.

Note. The RBGE copy has a printed label attached to p. 20, which gives an 'Explicatio Nominum Abbreviatorum'.

ANON. [HOPE, J.] (1778). *Catalogus arborum et fruticum in horto Edinensi crescentium anno 1778.* Edinburgi: Apud Balfour et Smellie. Pp. [i–ii], [1–]2–22 (lacks the 'Explicatio Nominum Abbreviatorum'). 8vo.

Note. Not previously noted, there were two editions of this work, with the same title page, but slightly different contents and pagination. Contemporary bills show that this work sold for one shilling.

ANON. [HOPE, J.] [c. 1779]. *Outlines of the history and progress of botany.* [Edinburgh]. Pp [1–]2–3.

Note. A copy of this previously unrecorded pamphlet (the 'heads' for Hope's lectures on the history of botany) has been located in the National Medical Library Bethesda, Maryland.

ANON. [HOPE. J.] [1780]. *Genera plantarum, ex editione decima tertia Systematis naturæ, illustrissimi Caroli a Linne. In usus academicos.* Edinburgi. Typis Academicis. Pp [i–ii], tab. 'Clavis Classium', [1–]2–106. 8vo.

Note. Contemporary bills show that this work sold for two shillings and sixpence.

HOPE, J. (1782). Beschryving van eene Budleja globosa. *Verhandelingen uitgegeeven door de Hollandse Maatschappye der Weetenshappen te Haarlem* 20(2): 417–8, t. 11.

HOPE, J. (1785). Description of a plant yielding Asa fœtida. In a letter from John Hope, M.D. F.R.S. to Sir Joseph Banks, Bart. P.R.S. *Philosophical transactions of the Royal Society* 75: 36–9, tt III, IV.

ANON. [HOPE, J.] [undated]. *A list of officinal plants.* [Edinburgh]. Pp [1–]2–8.

Note. Hope referred to this as 'Plantae officinales', one of his 'Books called Class books' of which there were two editions. A copy at the National Archives of Scotland with Hope's annotations (possibly for the second edition) suggests that the title was to be changed to 'Index herbarum medicinalium in horto Edinburgensi sub dio crescentium'.

A work in which Hope undoubtedly had an input is the Edinburgh *Pharmacopoeia*, produced by the Royal College of Physicians. The authorship is anonymous, but two editions (the 6th and 7th) were published in Hope's period:

ANON (1774). *Pharmacopoeia Collegii Regii Medicorum Edinburgensis.* Edinburgi, apud G. Drummond et J. Bell.

ANON (1783). *Pharmacopoeia Collegii Regii Medicorum Edinburgensis.* Edinburgi, apud Joannem Bell; et Londini, apud. Geo. Robinson.

Selected references

ANDERSON, W.P. (1931). *Silences that speak.* Edinburgh: Alex Brunton.

APPLEBY, J.H. (1983). 'Rhubarb' Mounsey and the Surinam toad: a Scottish physician-naturalist in Russia. *Archives of natural history.* 11: 137–152.

ARNOT, HUGO (1779). *The history of Edinburgh from the earliest accounts to the present time.* Edinburgh: William Creech.

[BALFOUR, I.B.] (1907). Eighteenth-century records of British plants. *Notes from the Royal Botanic Garden Edinburgh* 4: 123–192.

BONEY, A.D. (1988). *The lost gardens of Glasgow University.* London: Christopher Helm.

BOULGER, G.T. (1891). Hope, John (1725–1786) in *Dictionary of national biography* (ed. Sidney Lee) 27: 321–322. London: Smith, Elder, & Co.

CHALMERS, JOHN (ed.) (2010). *Andrew Duncan senior: physician of the Enlightenment.* Edinburgh: National Museums Scotland.

CRAIG, W.S. (1976). *History of the Royal College of Physicians of Edinburgh.* Oxford: Blackwell.

CRUFT, K. & FRASER, A. (1995). *James Craig 1744–1795: 'the ingenious architect of the New Town of Edinburgh'.* Edinburgh: Mercat Press.

DARWIN, F. (1909). A botanical physiologist of the eighteenth century. *Notes from the Royal Botanic Garden Edinburgh* 4: 241–4.

DUNCAN, A. (1789). *An account of the life, writings and character of Dr John Hope ... delivered as the Harveian Oration at Edinburgh 1788.* Edinburgh.

EDDY, M.D. (2003). The University of Edinburgh natural history class lists 1782–1800. *Archives of natural history* 30: 97–117.

EDDY, M.D. (2008). *The language of mineralogy: John Walker, chemistry and the Edinburgh Medical School 1750–1800.* Farnham: Ashgate Publishing Limited.

EMERSON, R.L. (1981). The Philosophical Society of Edinburgh: 1748–1768. *British journal of the history of science* 14: 133–176.

EMERSON, R.L. (1982). The Edinburgh Society for the importation of foreign seeds or plants 1764–1774. *Eighteenth-century life* 7 (n.s.): 73–95.

EMERSON, R.L. (1985). The Philosophical Society of Edinburgh: 1768–1783. *British journal of the history of science* 18: 255–303.

EMERSON, R.L. (1988). Lord Bute and the Scottish universities 1760–1792 in K.W. Schweizer (ed.) *Lord Bute: essays in re-interpretation*, pp 147–179. Leicester: Leicester University Press.

EMERSON, R.L. (2004). Select Society in *Oxford dictionary of national biography* (ed. H.C.G. Matthew and B. Harrison), 49: 705–8. Oxford: Oxford University Press.

FLETCHER, H.R. & BROWN, W.H. (1970). *The Royal Botanic Garden, Edinburgh 1670–1970.* Edinburgh: Her Majesty's Stationery Office.

GROVE, R.H. (1995). *Green imperialism: colonial expansion, tropical island Edens and the origins of environmentalism, 1600–1860.* Cambridge: Cambridge University Press.

HARVEY, J.H. (1981). A Scottish botanist in London in 1766. *Garden history* 9: 40–75.

HENDERSON, D.M. & DICKSON, J.H. (1994). *A Naturalist in the Highlands: James Robertson, his life and travels in Scotland 1767–1771.* Edinburgh: Scottish Academic Press.

HULTON, P., HEPPER, F.N. & FRIIS, I. (1991). *Luigi Balugani's drawings of African plants.* Rotterdam: A.A. Balkema.

KERR, R. (1811). *Memoirs of the life, writings, and correspondence of William Smellie ...* Edinburgh: printed for John Anderson; and Longman, Hurst, Rees, Orme and Brown.

KNIGHT, W. (1900). *Lord Monboddo and his contemporaries.* London: John Murray.

KOERNER, L. (1999). *Linnaeus: nature and nation*. Cambridge, Massachusetts: Harvard University Press.

MABBERLEY, D.J. (2004). Hope, John (1725–1786) in *Oxford dictionary of national biography* (ed. H.C.G. Matthew and B. Harrison) 28: 23–24. Oxford: Oxford University Press.

MCCARTHY, JAMES (2008). *Monkey puzzle man: Archibald Menzies, plant hunter*. Dunbeath: Whittles Publishing.

MILLER, D.P. (1988). 'My favourite studdys': Lord Bute as naturalist in K.W. Schweizer (ed.) *Lord Bute: essays in re-interpretation*, pp 213–239. Leicester: Leicester University Press.

MORTON, A.G. & NOBLE, M. (1983). Two hundred years of the biological sciences in Scotland: botany and mycology. *Proceedings of the Royal Society of Edinburgh*. 84b: 65–83.

OLIVER, F.W. (ed.) (1913). *Makers of British botany*. Cambridge: Cambridge University Press.

PATON, H. (1837–8). *A series of original portraits and caricature etchings by the late John Kay ... with biographical sketches and illustrative anecdotes*. 2 vols. Edinburgh: Hugh Paton.

PEEL-RITCHIE, R. (1899). *The early days of the Royal College of Physicians Edinburgh*. Edinburgh.

RISSE, G.B. (1986). *Hospital life in Enlightenment Scotland*. Cambridge: Cambridge University Press.

ROBINSON, T. (2008). *William Roxburgh: the founding father of Indian botany*. Chichester: Phillimore in association with the Royal Botanic Garden Edinburgh.

ROCK, J. (2000). An important Scottish anatomical publication rediscovered. *The book collector* 49: 27–60.

SCHWEIZER, K.W. (ED.) (1988). *Lord Bute: essays in re-interpretation*. Leicester: Leicester University Press.

SIMPSON, A.D.C. (2008). Thomas and John Donaldson and the Edinburgh medical class cards. Book of the Old Edinburgh Club 7(n.s.): 71–86.

SLACK, A.A. (1986). Lightfoot and the exploration of the Scottish flora, in *The long tradition, Botanical Society of the British Isles conference report No. 20* (ed. H.J. Noltie), pp 59–76; forming an un-numbered part of *The Scottish naturalist*.

SMITH, B.B. & FRANCOZ, C. (2009). *Botanic Cottage, Leith Walk, Edinburgh: building survey and historical research (Project 2627)*. Glasgow: GUARD (Glasgow University Archaeological Research Division).

Note. Copy of report (on behalf of Friends of Hopetoun Crescent Garden) held in RBGE Library.

SMITH, LADY [PLEASANCE] (1832). *Memoir and correspondence of the late Sir James Edward Smith, M.D. ...* 2 vols. London: printed for Longman, Rees, Orme, Brown, Green and Longman.

STAFLEU, F.A. (1971). *Linnaeus and the Linnaeans: the spreading of their ideas in systematic botany, 1735–1789*. Utrecht: A. Oosthoek for IAPT.

TAIT, A.A. (1980). *The landscape garden in Scotland 1735–1835*. Edinburgh: Edinburgh University Press.

TURNER, D.M. (1938). The economic rhubarbs: a historical survey of their cultivation in Britain. *Journal of the Royal Horticultural Society* 63: 355–370.

WALKER, MARGOT (1988). *Sir James Edward Smith M.D., F.R.S., P.L.S. 1759–1828: the first president of the Linnean Society of London*. London: Linnean Society of London.

WITHERS, C.W.J. (1991). The Rev. John Walker and the practice of natural history in late eighteenth-century Scotland. *Archives of natural history* 18: 201–220.

Acknowledgements

The reviser is greatly indebted to the following for their assistance with the preparation of this new edition:

First and foremost to Jane Corrie for her extensive work on the Hope archives, especially with reference to the Leith Walk garden, for her enthusiasm for the Botanic Cottage project, and for many helpful discussions. Also to Joe Rock for interesting conversations about art and architectural matters relating to Hope's garden and the artists he used.

Librarians and staff of the Special Collections of Aberdeen University (Michelle Gait) and Centre for Research Collections at Edinburgh University (especially Grant Butter, who found the Hope class ticket, and Dr Joseph Marshall), the National Library of Medicine, Bethesda, Maryland (Dr Stephen Greenberg), the Royal College of Physicians of Edinburgh (Iain Milne, who kindly photographed several of the portraits used here, and Estela Dukan) and the Linnean Society of London (Lynda Brooks, Gina Douglas, Ben Sherwood).

Dr Matthew Eddy (University of Durham) for his elegant paraphrase of Richard Sher's definition of the Enlightenment quoted in Chapter 1, for his information on the use of geometrical diagrams in the anatomy lectures of Monro *secundus*, and for conversations on other matters of mutual interest. Professor Charles W.J. Withers (University of Edinburgh) for useful discussions about Hope and related topics. Professor Christine Maggs (Queen's University, Belfast) for comments on the *Fucus* drawing.

Dr Patricia R. Andrew for confirming the attribution of the perspective views of the Leith Walk garden to Jacob More.

Stephen Astley of Sir John Soane's Museum for helpful discussions about, and sending images of, the Linnaeus monument designs from the office of Robert Adam.

Peter Wagner of Copenhagen for information on the Fabricius brothers. Professor John Parker for information on Thomas Martyn at Cambridge.

Andrew McLean and the private collection at Mount Stuart for providing, and permission to reproduce, the photograph of the Allan Ramsay portrait of Hope's Maecenas, the 3rd Earl of Bute.

Mr and Mrs Keith Adam for a wonderful visit to Blair Adam, for permission to reproduce the Francis Cotes portrait of Keith's ancestor John, for interesting discussions about Patrick Brydone and the Castagno dei Cento Cavalli, and for showing me the cottage John Adam designed for his own gardener, which belongs to the same natural order, if not the same genus, as the one he designed for his friend Hope's gardener. John McKenzie who photographed the Adam portrait.

The Friends of Hopetoun Crescent Garden (www.hopetouncrescentgarden.org.uk), Eileen Dickie, James Simpson and Tom Addyman. Diane Baptie for genealogical delvings. Last, but by no means least, RBGE staff in the following divisions: Herbarium (Adele Smith and Elspeth Haston for photographing the teaching drawings; David Mann for comments on the *Fucus* drawing, David Long and his 'Bryonet' contacts on the *Splachnum*, and Mark Watson), Library (Jane Hutcheon, Graham Hardy, Leonie Paterson), Exhibitions (Paul Nesbitt, Linsey Young) and Publications (Hamish Adamson and especially Caroline Muir for designing the book).

The Edinburgh Botanic Garden (Sibbald) Trust and Alisoun Morton for permission to tamper with the original text and the latter for donating her father's papers to the RBGE library and providing biographical information on him.

It is also only proper to repeat the names of those acknowledged in the first edition (though most of the institutions mentioned therein continued to give generous help towards the present work):

J.K. Burleigh W.S. (Drummond and Co.), Professor Roger L. Emerson (University of Western Ontario), † H. Heine (Paris), M.V. Mathew (RBGE), Anne McCulloch, Dr Athol L. Murray (Scottish Record Office), Jennifer Lamond (RBGE), †Dr B.C. Stone (Philadelphia). Staff of the National Library of Scotland.

To which must be added † B.L. Burtt and I.C. Hedge, whose names, from characteristic modesty, were omitted, but whose considerable background work over several years, and the task of seeing the work through the press, are apparent from Morton's and Burtt's personal papers.

Those marked † are known to have subsequently died.

Index